An Intervention Strategy Designed to Reduce Math Remediation Rates at Community Colleges is the property of:

Adventures In Education, Inc.
3460 S Fletcher Ave Ste 205
Fernandina Beach, FL 32034
eknowlton@adventures-in-education.com
http://www.adventures-in-education.com

AF680610

Copyright © 2015 by Adventures In Education, Inc.
All Rights Reserved. No part of this book may be reproduced or transmitted in any form or by any means, electronic or mechanical, including photocopying, recording, or by any information or retrieval system, without permission in writing from the published. The published hereby grants to individual teachers who have purchased this book permission to reproduce the activities and projects needed for use with their own students. Reproduction of the book for an entire school or school district or for commercial use is prohibited.

Printed in the United States of America.

An Intervention Strategy Designed to Reduce Math Remediation Rates at Community Colleges

Dissertation

Submitted to Northcentral University

Graduate Faculty of the School of Education
in Partial Fulfillment of the
Requirements for the Degree of

DOCTOR OF PHILOSOPHY

by

ELISABETH H. KNOWLTON

Prescott Valley, Arizona
March 2011

Abstract

In 2007, over 70% of students entering community college enrolled in developmental math. While the intent of assessment is to fairly place students in appropriate courses, the tools and strategies currently utilized do not produce the intended results. The purpose of this quasi-experimental quantitative study was to explore the efficacy of a new assessment intervention, namely remediating deficiencies identified via ALEKS Computerized Placement Test (CPT) and offering students the option to retake the CPT to improve placement outcomes. A one-group, pre-post design was used to examine the effect of the intervention on test scores, placement outcomes, and completion rates. A community college in the Southwest, the first college to utilize the ALEKS testing instrument for math assessment, review, retesting and placement purposes, served as the research site. Participation was limited to students who scored below the cutscore and volunteered to review material prior to retaking the CPT. A paired *t* test was used to compare pre- and post-CPT scores for 172 students and showed significant increases in test scores [$t(171) = 12.13, p < .0001, \alpha = .05$]. Group comparability among demographics was determined via ANCOVA and results showed significant test score increases for males [$F(1,151) = 10.43$, $p = .0015$] and Asian, White, and Hispanic students [$F(5,151) = 15.66, p < .0001$] at $\alpha = .05$. Overall 54% of students improved placement; 29% gained access to college-level math and 25% improved placement within remedial math. Results of one-sample *z* test for proportions ($z = -.3229, p = .6266, \alpha = .05$) indicated that students who placed in college-level math as a result of the intervention were as successful as those originally placed. Logistic regression results identified variable gender significant in predicting placement outcomes [$\chi^2(1) = 15.74, p < .0001$] and success rates [$\chi^2(1) = 14.69, p = .0001$] with male students more likely to improve placement and successfully complete the course ($\alpha = .05$). Further research is warranted to substantiate the efficacy of this new assessment placement approach and reliability of ALEKS. A holistic view on assessment placement would serve institutions well in promoting greater student success.

Acknowledgments

I extend my utmost gratitude to my exceptional dissertation committee. On numerous occasions, Dr. Rice, Dr. Hauer, Dr. Liu, and Dr. Pucci devoted a holiday weekend to reviewing my manuscript. I can't thank you enough for your dedication, responsiveness, and advice. Particularly, I am extremely grateful to Dr. Donna Rice, my committee chair, for her guidance, meaningful and uplifting feedback, and most of all her never failing promptness. I could not have completed my degree without your exemplary mentorship and support. Thank you!

I would be remiss if I did not thank Mr. Matt Ashcraft and his associates for his advice on the data retrieval process, and his assistance and cooperation in accessing data from the college database. Your time and efforts are greatly appreciated. I thank Dr. Bala Hosmane, my former professor at NIU, for availing the use of SAS software and his valuable statistical counsel. I also thank Dr. Alan Gorr, Dean of Education at Benedictine University, for planting the seed and encouraging me to reach for this ultimate goal.

And there are many others, professional associates, colleagues and personal friends, who have provided their support and encouragement during this time. I like to thank all of you! First and foremost, I am extremely grateful to my husband Tom. As I reached the final stage of this journey and reflected on the past four years, I realized that you were the one who made the fulfillment of my life-long goal possible. You encouraged me to take the leap to start this challenging endeavor, created an environment that allowed me to soar to new heights, and were always there to help me through difficult experiences. You knew that once I started this journey I would see it through. You sacrificed much time and provided the space I needed to work, study and sleep, and assumed many household and business duties while I was occupied studying. I can't thank you enough for the continual support and inspiration you provided along the way and your steadfast confidence in my ability to succeed. I am forever grateful to you and I deeply appreciate everything you have done. I could not have achieved this without you, and dedicate this dissertation to you Tom.

I also want to thank my son Greg for lending your support and understanding how important this endeavor was to me. What you didn't know is that I strove to complete my degree by the time you finished your baccalaureate degree in the hope of inspiring you to strive for excellence in your educational pursuit. You beat me to the punch, but I came close and our graduation was only six months apart.

My mother in Germany also played a major role in my success. Even though she didn't fully understand, she always lent her support and expressed great pride in my incremental achievements. Thank you Mom for your love, encouraging words, and long-distance prayers. Vielen, vielen Dank Mutti fuer deine liebevolle Unterstuetzung und zahlreichen Gebete!

Table of Contents

List of Tables

List of Figures

Chapter 1: Introduction

Much has been written regarding the remedial math dilemma at community colleges, namely the increasing need of entering college students for remediation and the controversial assessment placement practices. In 2007, over 70% of students entering community college enrolled in developmental math (Biswas, 2007; Brown & Niemi, 2007). More than half of the students who completed three years of college preparatory math and almost 25% of students who took calculus in high school tested into a remedial math course (Ruiz, 2007).

The unpreparedness of high school graduates to perform college-level work in mathematics is well established (Biswas, 2007; Brown & Niemi, 2007; Kirst, 1997; NCES, 2003; Smydo, 2008). A reflection of this trend is apparent in the vast number of remedial courses offered by higher education institutions in the United States, which continues to increase at an alarming rate. Many educators, policy makers, and the public advocate a need to reduce the number of students placed in remedial math and to increase degree completion rates.

Literature delineating possible causes of the increasing need for math remediation at the postsecondary level is quite extensive. Studies by Biswas (2007), Brown and Niemi (2007), and numerous other researchers identified assessment placement processes and testing procedures at community colleges key contributors to the dilemma (Collins, 2008; Golfin, Jordan, Hull, & Monya, 2005; Kirst, 2001; Mekonnen & Reznichenko, 2007). Assessment is considered a pivotal point in a student's college career and high stakes are attached to placement decisions. Accurate assessment of entering college students' math proficiencies and appropriate course placement are critical components of successful educational programs. Beginning courses at the right level is very important for students (Felder, Finney, & Kirst, 2007; ICHE, 2008; Kirst & Bracco, 2004; Morgan & Michaelides, 2005).

Placement tests are a source of controversy as they possibly exclude students rather than aid access (Kirst, 2001; Ruiz, 2007). Research has shown that placement in remedial math courses increases dropout rates, decreases course completion, delays student progress toward completion of a degree, and greatly reduces the probability of completing a degree (Biswas, 2007; Brown & Niemi, 2007; Felder et al., 2007; ICHE, 2008; Kirst & Bracco, 2004).

Studies by the National Center for Public Policy and Higher Education (Brown & Niemi, 2007), the Institute for Educational Leadership (Kirst, 2001), and the Achieving the Dream Foundation/Jobs for the Future (Biswas, 2007; Collins, 2008) delineated the magnitude of this problem and emphasized that colleges focus more on the initial assessment process to evaluate students and affect more correct placement. Improving assessment placement processes deserves high priority, and community college leaders are urged to seriously consider alternatives to remedy the situation (Brown & Niemi, 2007; National Center, 2008).

Background

Shults (2000) and Smydo (2008) remarked that assessment and placement of students into remedial courses remains one of the most debated aspects of remedial education. A large number of studies identify shortcomings in assessment placement processes and the accountability of test procedures used at community colleges as key contributors to the remedial math dilemma (ACT, 1993; Biswas, 2007; Brown & Niemi, 2007; Collins, 2008; Golfin et al., 2005; Kirst, 2001; MeKonnen & Reznichenko, 2007; Ruiz, 2007).

The increase in enrollment of entering college students considered to be underprepared has led to a growing trend at community colleges to implement a mandatory placement policy (Collins, 2008; Golfin et al., 2005; Prince, 2005; Ruiz, 2007; Shults, 2000). Almost every community college has some type of placement instrument to determine college readiness of entering students (Felder et al., 2007; Ternes, 2009). Since placement tests used by community colleges must accommodate the large influx of students with diverse educational backgrounds, standardized tests are the most economical choice, require little effort to implement, and are easily accessible and maintained (Ruiz, 2007).

A few commercially distributed CPTs dominate the field, namely ACCUPLACER, COMPASS and ASSET (Brown & Niemi, 2007; Felder et al., 2007; Gerlaugh, Thompson, Boylan, & Davis, 2007; Golfin et al., 2005; Kirst, 2001; Kirst & Bracco, 2004; Ruiz, 2007; Ternes, 2009), all of which are standardized norm-referenced tests (Brown & Niemi, 2007; Gerlaugh et al., 2007). Such tests are a source of controversy because of the small number of questions administered, namely 7 to 17. Research has substantiated that only a sufficiently large number of questions, e.g. more than 20 questions, will accurately assess broad content area and raises the question with respect to validity and reliability of most commonly used assessment tests (Mekonnen & Reznichenko, 2007). Other research findings attributed the large number of students placed in developmental courses to the unfamiliarity of students with the type and content of the test, the time elapsed since graduation, and the lack of preparation (Hager, 2005; Kirst, 2003; Ruiz; Thompson, 2006).

To improve student outcomes in the testing process, some institutions have created preparation workshops and practice exams to better prepare students for standardized test taking and familiarize students with the test content (Mekonnen & Reznichenko, 2007; Ruiz, 2007). A study by Ruiz (2007) found statistically significant improvements in standardized test scores as a result of preparation workshops. Research by Golfin et al. (2005) substantiated that providing applicants with a practice exam before the actual placement test reduced the number of students required to take developmental math.

Various other studies suggested that consideration of a new approach to remediation, namely identifying and remediating deficiencies, should be given in particular for students who had learned math in their secondary education (Abraham & Creech, 2000; Biswas, 2007; Davis, 1999; Mekonnen & Reznichenko, 2007; Ruiz, 2007; Thompson, 2006). As most testing instruments, e.g. ACCUPLACER, COMPASS and ASSET, are imperfect for assessing student strengths and weaknesses (Kirst, 1997; Prince, 2005), students who fail the college placement test are only provided a numerical score and no specific information on what to review or classes to address their deficiencies (Kirst).

Another approach, namely offering students the option to retake the placement test, has received some consideration by community colleges (Calcagno & Long, 2008). Scant research on the efficacy of retake CPTs is available. A study by Geraci (2007) explored the efficacy of the retake option on student outcomes. Geraci concluded that allowing students to retest improved retention rates and resulted in a modest increase in scores, in particular for those students whose initial score was close to college level placement.

While community colleges make a concerted effort to improve remediation rates and experiment with various approaches to remedy the situation, a synergistic approach to assessment placement appears to be lacking. Shortcomings in assessment placement beckon a more comprehensive approach to affect more correct placement and improve student outcomes. A program designed by a community college in the Southwest and piloted in 2010 presented an innovative approach to assessment placement. The program used a new testing instrument developed in 2009, namely ALEKS CPT and Learning Modules (ALEKS, 2010). The ALEKS test is used for assessment, review, retesting and placement purposes.

The instrument inherits a groundbreaking design in Knowledge Space Theory and assesses student proficiencies, performs an in-depth analysis of student content knowledge, and provides an individualized detailed report listing topics in need of remediation (Falmagne, Cosyn, Doignon, &Thiery, 2003). ALEKS Learning Modules, a remediation tool, offers students a medium for targeted review. ALEKS system, a standardized criterion-referenced test (CRT), is fully automated and adaptive incorporating 20-35 open response questions. Because of its adaptive mechanism the system also lends itself for retesting purposes. The college identified and given the fictional name Southwest Community College (SWCC) served as a testing ground for investigating the efficacy of incorporating targeted math review and retake CPT option into the assessment placement process to affect more correct placement, reduce remediation rates, and improve student success rates.

Problem Statement

Research has shown that remedial instruction delays a student's progress toward completion of a degree, greatly reduces the probability of completing a degree, and increases cost for students, parents, institutions, and taxpayers (Brown & Niemi, 2007; Kirst & Bracco, 2004). According to a 2000 study by the Mackinac Center for Public Policy remedial education taxes the United States economy at a rate of 16.6 billion dollars per year (Parr, Edwards, & Leising, 2006). A 2006 report by the nonprofit Alliance for Excellent Education revealed that community colleges alone spend $1.4 billion annually on remedial courses for recent high school graduates (Smydo, 2008). "Taxpayers are essentially paying twice for the course work and skill development students are expected to receive in high school" (Smydo, 2008, para. 18).

A new approach to math assessment placement for entering community college students deemed warranted. Numerous researchers suggested that colleges focus on identifying and remediating specific deficiencies for students who had learned higher level math in their secondary education, e.g. targeted remediation, and offering students the opportunity to retake the placement exam (Abraham & Creech, 2000; Biswas, 2007; Calcagno & Long, 2008; Davis, 1999; MeKonnen & Reznichenko, 2007; Ruiz, 2007; Thompson, 2006). Currently, students who fail the computerized placement test (CPT) are not provided specific information on what to review to address their deficiencies (Kirst, 1997) nor offered the opportunity to retake the CPT (Calcagno & Long, 2008; Geraci, 2007).

Most placement testing instruments are imperfect for assessing student strengths and weaknesses (Kirst, 1997; Prince, 2005). Results of one-on-one interviews with students, who successfully completed college preparatory math courses in high school, revealed that many students were disappointed with their placement. Students claimed they could achieve higher placement if they were provided a list of topics to review (Ruiz, 2007). Few research studies have been completed on the efficacy of retake policies (Calcagno & Long, 2008; Geraci, 2007). Virtually no attention has been devoted to identifying specific deficiencies and integrating targeted remediation into the assessment placement process to improve placement outcomes and reduce remediation rates (Abraham & Creech, 2000; Biswas, 2007; Prince, 2005; Thompson, 2006).

Purpose

The purpose of this quasi-experimental one-group pre-post quantitative study was to explore the efficacy of implementing targeted math remediation into the assessment placement process to improve test scores, placement outcomes, and success rates of entering community college students. A quantitative methodology best met the needs and purpose of the research study as it allowed investigation of the relationship between targeted math remediation, the independent variable, and student outcomes, the dependent variable (Larson & Farber, 2003). As such it helped provide a better understanding of how to reduce remediation rates as a result of more accurate placement, and, in turn, possibly improve college success and degree completion rates.

Participation in the study was limited to students who did not meet the requirements to enroll in college-level math, i.e. scored below the cutscore, and volunteered to review material prior to retaking the placement test. The cutscore was used to identify students ineligible to enroll in college-level math and thus eligible to participate in the program. Students self-selected to participate in the program through guidance of campus personnel, e.g. testing personnel and counselors.

Prospective power analysis via G*Power analysis software with power set at .80 and α = .05 assuming a medium effect size, indicated a required sample size of 90 for a two-tailed *t* test with two dependent samples and a sample size of 150 for ANOVA. A sample size of 172 sufficed to validate results. See discussion in Ch 3 Limitations on requirements for logistic regression sample size.

Theoretical Framework

"The underlying philosophical issue that [drove] the research endeavor" (Trochim & Donnelly, 2007, p. 4) was the promotion of equitable systems, an advocacy/participatory worldview (Lather, 2003). In this case the possible misplacement of students capable of success in college-level math but placed into remedial math courses. The advocacy/participatory worldview holds that research advocates for an action agenda that helps groups or individuals that may be marginalized (Kemmis, 2006), in this case students who were denied the opportunity to attempt college-level math courses and as such an issue that deserved attention as the present structure limited students' self-development.

Paulo Freire, a critical theorist, noted that a critical theory approach of empowering research can be instrumental in raising consciousness for the need to change and for advancing an agenda to improve lives, e.g. students' educational attainment (Lather, 2003). Freirian research "focuses on promoting liberation and growth within a society as assumed to be class divided, and hence inequitable" (Lather, 2003, p. 199). Low-income and minority students are traditionally overrepresented in community college developmental education programs (Prince, 2006; Ruiz, 2007). The proposed intervention paralleled the concepts characterized in Freirian research (Lather, 2003) as it created opportunities for students to be liberated from possible misplacement and availed opportunity for educational advancement, i.e. growth. Bohman (2005), in a publication in Stanford Encyclopedia of Philosophy, stated that a critical theory approach is adequate only if it meets three criteria: (a) explains a problem in current social reality, (b) identifies the actors to change the problem, and (c) has achievable and practical goals for social transformation.

The current social reality across U.S. community college campuses indicated that a number of students may be placed into developmental math classes who could be successful in college-level courses (Armstrong, 2000; Ruiz, 2007). A wealth of literature on remedial math education exists with much of the literature focusing on pedagogical concerns and approaches, and most of the studies neglecting to address identification and remediation of specific content areas (Parmer & Cutler, 2007).

The proposed quantitative study investigated the factors contributing to the remedial math dilemma. The phenomenon is of great complexity and literature on remedial math, in particular on math assessment placement processes and practices, is quite extensive (Biswas, 2007; Brown & Niemi, 2007; Kirst, 2003; Kirst & Bracco, 2004; Levin & Calcagno, 2008; Ruiz, 2007; Thompson, 2006; Venezia, Kirst, & Antonio, 2003). Review of literature revealed numerous factors (independent and intervening variables) influencing student success (dependent variables) and allowed identification of two key dependent variables, namely college placement test (CPT) scores and placement in remedial math (Armstrong, 2000; Brown & Niemi, 2007; Cozby, 2009; Golfin et al., 2005; MeKonnen & Reznichenko, 2007; Ruiz, 2007; Thompson, 2006). The theoretical framework shown in Figure 1 portrays the relationship among the factors identified and follows the design introduced by Jungnickel and Creswell (1994).

The causal model constructed depicts the possible causal sequence among variables and aids in visualizing the interconnections of variables. The exogenous variables are listed on the far left, the endogenous variables in the middle, and the dependent variables on the right. The direction of influence flows from left to right with plus and minus signs indicating the hypothesized direction, e.g. a plus sign in the direction of CPT scores indicates the likelihood of an increase in CPT scores, a plus sign in the direction of placement in remedial math indicates a possible increase for the need of remedial math, etc. For this particular quantitative study, fifteen exogenous variables influence the two identified dependent variables, mediated by the influence of three endogenous variables.

Applied to the proposed quantitative study, critical theory holds in that introducing new elements into the assessment placement and remediation process, the independent variables, could possibly explain an increase in CPT scores and accelerate pathway through remediation, the dependent variables (Cozby, 2009). Advancing student proficiencies could also possibly redirect student behavior and motivation, endogenous variables, to achieve higher levels of success (Marzano & Kendall, 2007). Although student motivation and time allotted to remediate deficiencies most likely influenced student performance, these variables were not actually measured in the study. Improving student proficiencies through targeted remediation during the assessment placement process most likely increased CPT scores and expedited the remediation process (Parmer & Cutler, 2007).

The proposed study met the three criteria of a critical theory approach stipulated by Bohman (2005) as it explained a problem in current social reality, identified the actors to change the problem, and had achievable and practical goals for social transformation. The goal of this study was to establish a possible framework for incorporating intervention strategies into the assessment placement and remediation process to positively impact student placement. The issue may not be localized to a single community college district and positive results of this research could be duplicated at other districts. Furthermore, improving college opportunities and providing access to higher education, increases the nation's ability to advance educational attainment of our workforce and citizenry and ultimately increase global competition (Brown & Niemi, 2007; National Center, 2008).

Figure 1. A Visual Model of a Theory of Remedial Math at Community Colleges

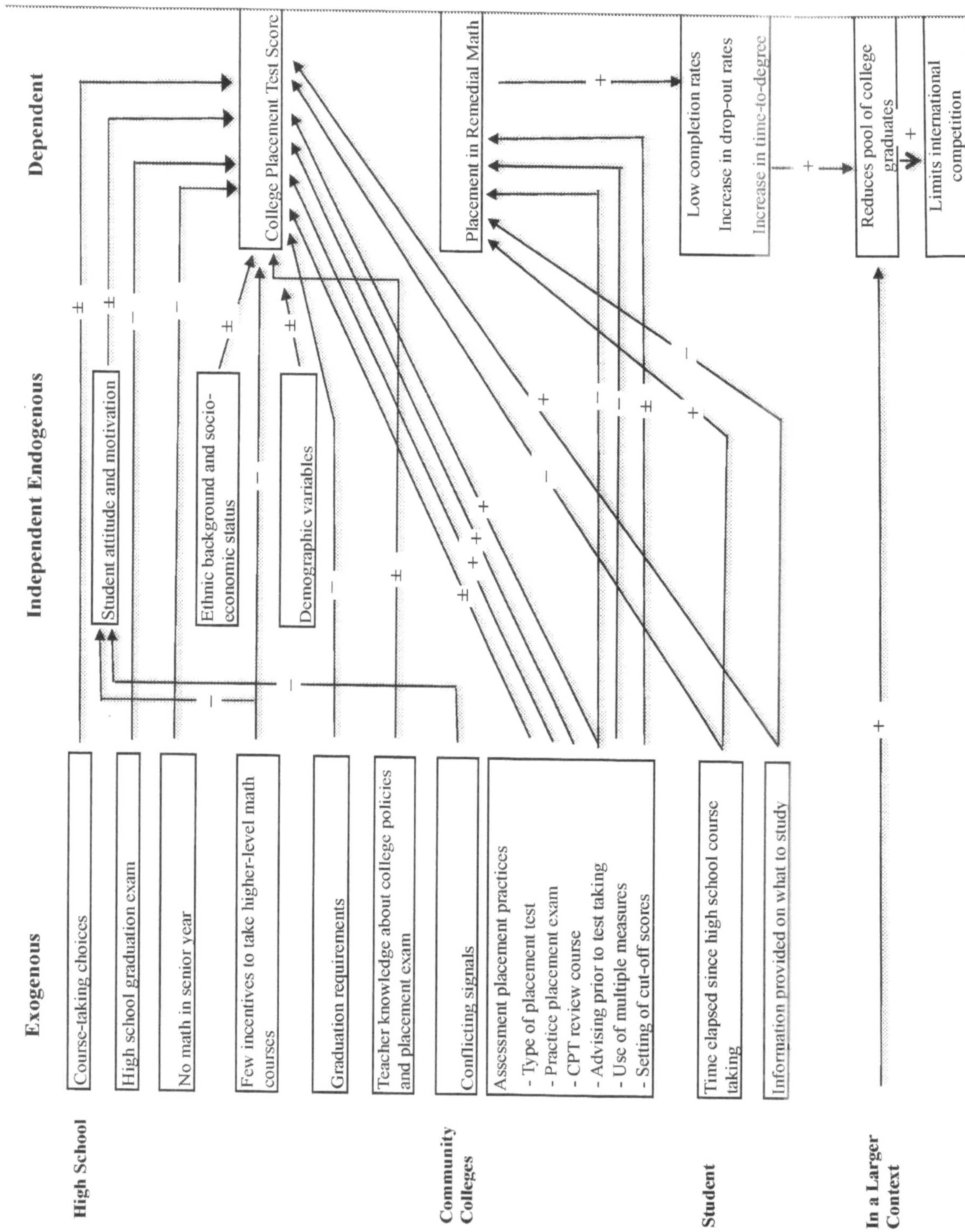

Figure 1 A Visual Model of a Theory of Remedial Math at Community Colleges

Research Questions and Hypotheses

The purpose of this quasi-experimental quantitative research was to investigate the efficacy of integrating a math assessment placement intervention into the assessment placement process, namely targeted math remediation. The proposed intervention offered students the opportunity to remediate specific math deficiencies identified via the initial CPT and retake the CPT to improve placement outcomes. The following research questions guided the proposed study.

Q1. To what extent did targeted math remediation during the assessment placement process affect CPT scores?

H1$_0$: There was no statistically significant difference in student initial and retake CPT scores for students who engaged in targeted math remediation.

H1$_a$: There was a statistically significant difference in student initial and retake CPT scores for students who engaged in targeted math remediation.

Q2. To what extent did targeted math remediation during the assessment placement process affect student placement?

H2$_0$: There was no statistically significant difference in student placement outcomes for students who engaged in targeted math remediation.

H2$_a$: There was a statistically significant difference in student placement outcomes for students who engaged in targeted math remediation.

Q3. To what extent did targeted math remediation during the assessment placement process affect course completion rates?

H3$_0$: There was no statistically significant difference in student success rates for students who placed in college-level math as a result of the intervention compared to students who originally placed in the course.

$H3_a$: There was a statistically significant difference in student success rates for students who placed in college-level math as a result of the intervention compared to students who originally placed in the course.

These questions were derived from the possible misplacement of students into remedial math at community colleges, in particular those students who had previously learned higher levels of math (Golfin et al., 2005; MeKonnen & Reznichenko, 2007; Ruiz, 2007). The objective of this research was to investigate a possible intervention strategy to reduce the number of students placed in remedial math. The final aspect of this research was to understand if targeted remediation during the assessment placement process would help students reach higher levels of success.

Nature of the Study

A quasi-experimental one-group pre-post quantitative design facilitated a methodology which best met the needs and purpose of the research in studying the efficacy of the proposed intervention, a cause-effect relationship (Trochim & Donnelly, 2007). This approach allowed investigation of the relationship between targeted review, the independent variable, and student outcomes, the dependent variable. The basic design followed a repeated-measures design implying that the same measure, i.e. CPT, was administered before and after the intervention (Coladarci, Cobb, Minium, & Clarke, 2008).

Significance of the Study

Math placement and student progress in remedial math could be improved through the addition of more data on incorporating new elements into the assessment placement process at community colleges. Results addressed the widespread problem of math remediation at community colleges and revealed possible avenues to improve assessment placement practices, increase placement test scores and decrease remediation rates, while at the same time possibly reduce dropout rates and increase completion rates. This topic of research appealed to a broad audience and is of national interest. Improvements in the remediation and assessment placement process would not only benefit students and parents, but also institutions, teachers, taxpayers, and the nation (Brown & Niemi, 2007; Kirst & Venezia, 2006; Levin & Calcagno, 2008). The nation's poor economic conditions and increased unemployment rates, coupled with decreases in local, state and government funding, diminishing institutional budgets, and increasing global competition, underscored the urgency to address the inherent problems in developmental mathematics at community colleges.

Definitions

Unique terms were critical to an understanding of the proposed study. This section provides definitions of such terms. The definitions are commonly used definitions within the body of relevant literature.

ACT. ACT is an acronym for American College Testing program. The ACT test is a globally recognized college admission test and one of the most widely accepted college entrance exams. The test assesses students' high school educational development and their ability to complete college-level work (ACT, 2010).

ALEKS. ALEKS, an acronym for Assessment and LEarning in Knowledge Spaces, is a web-based artificially intelligent assessment and learning system developed by ALEKS corporation. The ALEKS system is fully automated incorporating open response (not multiple choice) and adaptive assessment (ALEKS, 2010).

ALEKS Math Placement Test. ALEKS Math Placement Test is a computerized criterion-referenced placement test. The test assesses student math proficiencies, performs an in-depth analysis of student content knowledge, and provides an individualized detailed report listing topics in need of remediation. The placement test incorporates two separate tests to accurately assess student current math knowledge and to provide appropriate review topics. ALEKS Level I Math Test can place students into Basic Arithmetic, Introductory Algebra, and Intermediate Algebra courses. ALEKS Level II Math Test can place students into courses such as Intermediate Algebra, College Mathematics, College Algebra/Functions and above (MCCCD, 2010). For the purpose of this study only Level II Math test results will be used.

ALEKS Learning Modules. ALEKS learning modules provide a medium that facilitate independent review and accommodate individual students' learning needs. Students who do not obtain the required minimum score for the math course they desire to take have the option to review topics identified on the initial ALEKS Math Placement Test via ALEKS learning modules and retake the test (ALEKS, 2010; MCCCD, 2010).

College-level math courses. The following courses are considered college-level math at SWCC: College Mathematics (MAT140, 141, 142), College Algebra/Functions (MAT150, 151, 152), Math for Elementary Ed (MAT156), Finite Math (MAT172), Trigonometry (MAT182), Precalculus (MAT187), Statistics (MAT206), and Brief Calculus (MAT212) (MCCCD, 2010).

Completion rate/success rate. Completion rate is the proportion of students who successfully completed a course compared to the total number of students enrolled, and does not include students who withdrew or received an incomplete (MCCCD, 2010).

Computerized Placement Test (CPT). The CPT is used for placement when students cannot provide recent SAT or ACT scores. Most commonly used CPTs are ACCUPLACER developed by the College Board (College Board, 2009b), and COMPASS and ASSETT both developed by ACT (ACT, 2010). ALEKS is a new CPT developed by ALEKS Corporation in 2009 (ALEKS, 2010).

Cutscore or cutoff score. A pivotal point that determines the placement of a student based on the placement test score received. A test score above the cutscore will place a student into a higher math class than a test score below the cutscore (Ruiz, 2007).

Deficiencies in mathematics. Deficiencies may be removed by scoring above the cutscore set for college-level math on the assessment test used for placement or completing developmental/remedial coursework with a grade C or higher (MCCCD, 2010).

GPA. GPA is an acronym for grade point average, a weighted average that indicates the student's overall performance (MCCCD, 2010).

High-stakes test. A high-stakes test is a test in which decisions about an individual hinge on the outcome of the test (Ruiz, 2007).

Math placement test. A test used to determine students' math skills and used to place incoming freshmen into classes congruent with their skill level (Geraci, 2007; Ruiz, 2007).

Remedial/Developmental math courses. Remedial courses are courses below the 100 level. The following courses are considered remedial math courses at SWCC: Arithmetic Review, Beginning Algebra, and Math Concepts and Applications (MCCCD, 2010).

Retake ALEKS Math Placement Test. Students who do not obtain the required minimum score for the math course they desire to take have the option to retake the ALEKS Math Placement Test. Review of topics identified as deficient on the initial test is strongly suggested prior to retaking the test. The highest score will be used for placement. Scores are valid for one year prior to the first day of enrollment in a math course. Students can only take two ALEKS Placement Tests per year. The college will pay for two math placement tests (MCCCD, 2010).

Retention. Retention is the ability to keep students enrolled in a program until graduation or course/competency completion (MCCCD, 2010).

SAT. SAT is an acronym for Scholastic Aptitude Test administered by The College Board. The SAT test is a globally recognized college admission test and one of the most widely accepted college entrance exams. The test assesses students' high school educational development and their ability to complete college-level work (College Board, 2009).

Success. Success of a student's academic achievement in a course will be defined as a final grade of C or higher (Ruiz, 2007).

Transfer-level courses. Transfer-level courses are college-level courses and transfer to a four-year institution (Ruiz, 2007).

Underprepared students/underperforming students/developmental students/remedial students. These classifications refer to students who lack basic skills required for success in college-level academics (Ruiz, 2007).

Summary

Studies have shown that current trends in remedial math education, i.e. the continual increase in the number of entering college students in need of remediation, impede student achievement (Biswas, 2007; Brown & Niemi, 2007; Felder et al., 2007; ICHE, 2008; Kirst and Venezia, 2004). A report by the National Center for Public Policy and Higher Education stressed that community colleges make a concerted effort to reduce the number of students placed in remedial math and explore avenues to reverse this trend (Brown & Niemi, 2007). Numerous research studies delineated the assessment placement process one of the key contributors to the dilemma as it appears that a significant number of students, in particularly those on the edge of remediation, may be unfairly denied the chance to attempt college-level courses due to shortcomings in current assessment placement practices (Collins, 2008; Golfin et al, 2005; Kirst & Bracco, 2004; Mekonnen & Reznichenko, 2007).

In an effort to affect more correct placement and improve student outcomes, many community colleges have experimented with various new approaches to assessment placement, including SWCC. The innovative design implemented by SWCC was of particular interest in this study as it not only incorporated the new placement testing instrument ALEKS, but also offered targeted review and retake CPT option. The proposed study was designed to examine the efficacy of the new assessment placement program designed by SWCC. The purpose of this research was to understand if targeted remediation during the assessment placement process and the option to retake the placement test would help students reach higher levels of success.

Chapter 2: Literature Review

The unpreparedness of U.S. high school graduates to do college-level work in mathematics is well established (Biswas, 2007; Brown & Niemi, 2007; NCES, 2003; Parr et al., 2006; Roth, Crans, Carter, Ariet, & Resnick, 2001; Smydo, 2008). A reflection of this trend is apparent in the vast number of remedial courses offered by higher education institutions in the U.S., which continues to increase at an alarming rate (Brown & Niemi, 2007; Parr et al., 2006; NCES, 2003; Smydo, 2008). In 2000, the National Center for Education Statistics determined that 76% of all institutions that admit freshmen offered at least one remedial math course, and 80% of public four-year universities and 98% of two-year public colleges offered such courses (Parr et al., 2006).

According to the Bridge Project, a national policy research study by Stanford University, over 70% of public high school graduates in the U.S. go on to postsecondary education within two years of graduating from high school (Kirst & Bracco, 2004). Over 50% of those students will take remedial courses with a large percentage not attending a second year of college, and 41% never completing a two- or four-year degree (Kirst & Bracco, 2004; Thompson, 2006). Research has shown that remedial instruction delays a student's progress toward completion of a degree and greatly reduces the probability of completing a degree (Brown & Niemi, 2007; Kirst & Bracco, 2004; Parr et al., 2006).

In addition, remedial education taxes the U.S. economy at a rate of 16.6 billion dollars per year according to a 2000 study by the Mackinac Center for Public Policy (Parr et al., 2006). A 2006 report by the nonprofit Alliance for Excellent Education revealed that, community colleges alone spent $1.4 billion annually on remedial courses for recent high school graduates

(Smydo, 2008). "Taxpayers are essentially paying twice for the course work and skill development students are expected to receive in high school" (Smydo, 2008, para. 18).

Many states across the nation are highly agitated by the academic problem that college freshmen experience (Biswas, 2007; Brown & Niemi, 2007; Gerlaugh et al., 2007; Kirst, 1997; NCES, 2003; Smydo, 2008; Venezia, 2000). The question remains what are the causes of remediation, non-completion, and inadequate secondary preparation. No one has accepted responsibility for this troubling trend, but instead engage in the blame game (Kirst, 1997; Smydo, 2008).

This literature review presents current research and publications regarding the perspectives on and explanations for the increasing need of remedial math education at the postsecondary level. The discussion leads to an investigation of current assessment placement practices and the various approaches taken by community colleges to remedy the situation. Because of the complexity of the dilemma, the final section will address remedial education in a larger context.

Remedial Math Dilemma

Who is to blame? Kirst (1997) remarked that, "Universities blame high schools for low standards; secondary schools blame elementary school; and elementary schools blame parental neglect" (para. 2). On the other hand, high school students and parents blame the educational system (Brown & Niemi, 2007; NCES, 2003; Smydo, 2008). The question that should be asked is what are the possible causes of the deterioration of student achievement and how can such be rectified.

The discussion presented in this section focused on identifying possible causes of the remedial math dilemma from the perspective of high schools, community colleges, students, and K-16 policy mechanism.

High schools. At the high school level, researchers identified numerous causes for the deterioration of student achievement and subsequent need for remediation. Such include but are not limited to: (a) Course-taking choices, (b) state high school graduation exams, (c) the math gap phenomenon, (d) teacher signals, (e) lack of incentives, and (f) focus on high school graduation rates. A discussion of each of the possible causes is provided below.

Course-taking choices. Critics of public education frequently interpret high failure rates and proliferation of postsecondary developmental courses as evidence that secondary school curricula and standards have become unacceptably diluted (Roth et al., 2001; Smydo, 2008). Numerous researchers claimed that, in many cases, course-taking choices in high school do not equip students with the skills needed to do college-level work (Biswas, 2007; Brown & Niemi, 2007; Kirst, 2003; Kirst & Bracco, 2004; NCES, 2003; Smydo, 2008; Roth et al., 2001) as high school curricula and postsecondary standards are not always consistent with one another (CSU, 2009; Kirst, 1998; Kirst & Bracco, 2004; Roth et al., 2001).

Furthermore, secondary students know they will be admitted to broad access institutions if they meet minimum GPA and course requirements, so they often take few academic courses, especially math, in their senior year (Kirst, 1998; Kirst & Bracco, 2004). A national survey conducted by ACT substantiated such as "40 percent of [high school students] bound for two-year college were not planning to take all the courses that ACT had deemed necessary for college level work" (Kirst & Bracco, 2004, p. 17). Many students instead opted to achieve high grades through taking undemanding courses and boost their GPA (Roth et al., 2001).

State high school graduation exams. State high school graduation exams are most often not aligned with tests used for college admission or placement in college-level courses (Brown & Niemi, 2007; Kirst, 2003; Kirst & Bracco, 2004; Thompson, 2006). By and large, high school tests are much easier covering content often not exceeding the ninth- or tenth-grade level (Kirst, 2003; Kirst & Bracco, 2004). Results of a content analysis of high school exit and college entrance tests conducted by RAND demonstrated significant differences in the content between assessments used in postsecondary compared to secondary school (Kirst & Bracco, 2004). As such these differences send mixed signals to students about college standards and preparation (Brown & Niemi, 2007; Kirst & Bracco, 2004). In 2000, data from the National Association of System Heads (NASH) study showed that only two states had aligned math requirements (Kirst & Bracco, 2004). A "most recent *Quality Counts* report from *Education Week* [indicated] that still only 11 states have aligned their state high school assessments with postsecondary education" (Brown & Niemi, 2007, p. 9; Edwards, Olson, & Swanson, 2007).

Math gap. Many students often do not take math during their senior year in high school since they have completed their state mathematics graduation requirement in their junior year (Blair, 2008; Kirst, 2003; Kirst & Bracco, 2004; Thompson, 2006). Due to the time elapsed since graduation they forget what they have learned and are caught in the "if you don't use it, you lose it" trap (Blair, 2008, para. 2). This phenomenon has been termed as the "math gap" (Thompson, 2006, para. 7). It is supported by a 2008 statistics, which revealed that almost 57% of ACT-tested students going directly to community college from high school were not ready for college-level mathematics coursework (Blair, 2008; Thompson, 2006). This gap can be particularly problematic for students when faced with taking a college entrance exam (Blair, 2008; Kirst & Bracco, 2004; Thompson, 2006).

Students who do not enter college immediately after high school also fall into this trap (Blair, 2008; Thompson, 2006). In 1996, Davis (1999) noted that nearly 86% of Illinois community college students enrolled in developmental education were 20 years of age or older. Most of the students mastered math in their secondary education, but the passage of time since high school, often exceeding 20 or more years, probably caused them to forget much of what they previously had learned (Davis, 1999).

Teacher signals. The Bridge Project found that school site educators can be purveyors of information, i.e. signals, about what students need to know in order to succeed at postsecondary education (Kirst & Bracco, 2004). Many teachers play a large role in providing such signals (Kirst & Bracco, 2004). However, high school teachers are not acquainted with community college placement tests, and they have no way of learning about their students' results on placement tests (Brown & Niemi, 2007; Kirst, 1997, 2003). Thus, it is difficult for teachers to effectively coordinate course content with university curricular expectations (Kirst, 1997). In addition, high school teachers and college professors often disagree on what knowledge and skills are required for postsecondary education (Kirst & Bracco, 2004).

Lack of incentives. Since it is so easy to enter most postsecondary schools, most of which are minimally selective or nonselective, there are few incentives for students to work hard in high school (Kirst, 2003; Kirst & Bracco, 2004; Ruiz, 2007). Many students believe there is little penalty for poor high school performance and that they will obtain a second chance at open enrollment institutions like community colleges (Kirst, 2003; Kirst & Bracco, 2004). As previously noted, ACT found that 40% of community college bound students did not take the ACT recommended coursework necessary for college (Kirst & Bracco, 2004). Kirst and Bracco (2004) remarked that academic disengagement is at an all-time high and seniors report spending less time on studying and homework than ever before. At the same time, senior-year high school grades continue to increase. Kirst and Bracco concluded that the combination of academic disengagement and record grade inflation pose a real challenge for our postsecondary education system.

Focus on high school graduation rates. Traditionally, the American education system has focused on providing opportunities for students to graduate, i.e. receive a high school diploma, and valued participation in the system for as long as possible (Kirst & Bracco, 2004). Getting a high school diploma has typically been more important than what was learned and schools traditionally received funding based on student seat time (Kirst & Bracco, 2004). A current wave of reform tries to address this issue, i.e. focus on knowledge and skills attained rather than receiving a high school diploma (Kirst & Bracco, 2004; National Center, 2008; Smydo, 2008).

Community colleges. At the community college level, researchers attributed the prevalence of remediation to low completion rates, course design, assessment placement practices, and conflicting signals. A discussion of each of these developments is provided below.

Low completion rates. "National longitudinal data show that at the end of five years, only 30% of community college students complete the associate degree" (Roth et al, 2001, p. 2). This low completion rate has caused concerns by critics that unprepared high school students are not being effectively remediated in community colleges (Roth et al., 2001; Smydo, 2008). The most common explanation offered by community college administrators for the high failure rate on the CPT is that students' course-taking choices in high school ill-prepared them for college-level work (Roth et al., 2001; Smydo, 2008). Community colleges further claim that it is not feasible to correct major academic deficiencies with one remedial course (Roth et al., 2001).

Course design. The Community College Research Center at Columbia University found troubling patterns in developmental math education as it analyzed the progress of students enrolled in Achieving the Dream institutions (Biswas, 2007). The study identified that at least part of the problem resides in the structure and delivery of developmental education (Biswas, 2007). "To address the needs of different students and to avoid trapping more students than necessary in developmental [math] courses," community colleges have begun experimenting with alternative delivery and design that allow students the option of a self-paced or accelerated pathway through developmental coursework (Biswas, 2007, p.1).

These flexible delivery models hold the promise of improved student outcomes since developmental education students enter at various levels of preparedness (Biswas, 2007). Some need minimal remediation, a refresher course, before they advance to college-level courses. Others are so academically underprepared that they require multiple levels of developmental courses before they are ready for college (Biswas, 2007; Smydo, 2008).

Assessment and placement practices. This section avails an examination of assessment placement practices at community colleges in terms of standardization and efficacy. A more in-depth discussion of assessment placement instruments, policies and practices is proved at a later point.

Lack of standardization. "There is currently no nationwide measure of college readiness or conceptual consensus on what credit-level, non-remedial postsecondary work requires" (Kirst, 2003, p. 8). No uniform placement testing process exists throughout the U.S. community college system (Brown & Niemi, 2007; Kirst, 2003). Few states have established cutoff scores for referral to developmental education and most have left such decisions to institutional discretion (Prince, 2005).

Placement exams are numerous, varied and largely unknown, and are designed to serve the assessment needs of departments within colleges and universities (Gerlaugh et al., 2007; Kirst, 2003). These exams are extremely diverse with many institutions using academic departmental faculties to devise a local exam (Gerlaugh et al., 2007; Kirst, 2003). They are not well publicized to high school students and their parents or to teachers, nor are they aligned with current K-12 standards and assessments (Venezia, 2000). None of the placement tests are designed to diagnose student math weaknesses, i.e. deficiencies in a particular topic within a course, so that high school students or entering college freshmen could better prepare for admission (Kirst, 1997). A few states like Texas have a statewide exam, but Texas colleges also use their own placement exams (Kirst, 2003).

In 1992, there were nearly 125 combinations of 75 different placement tests devised by universities with minimal regard to secondary school standards in the southeastern United States (Kirst, 2003). California community colleges use nearly over 60 different placement exams with 678 combinations of exams (Kirst, 2003; Venezia, 2000). The difference between senior high school and postsecondary assessment is large and reflects a tremendous variability of placement testing practices (Brown & Niemi, 2007; Kirst, 2003). "Such variety yields inconsistency throughout the community college system not in only test content but also in levels of expected proficiency within a given subject area domain" (Brown & Niemi, 2007, p. 3).

Efficacy of placement testing. Research on the efficacy of placement testing has produced mixed results and stimulated great controversy (Brown & Niemi, 2007; Ruiz, 2007). A review of research on the predictive validity of placement tests found that, "most of the studies relating community college placement test scores to course grades resulted in small or modest relationships" with low or moderate validity coefficients (Brown & Niemi, 2007, p. 5). Various studies pointed to repeat findings that ineligible students (students identified through placement exams as unprepared or ineligible for credit-bearing courses) were generally 70-80% as successful as eligible students (Armstrong, 1994; Brown & Niemi, 2007; Ruiz, 2007). This strongly suggests that students may be unfairly denied the chance to attempt a college-level course despite their capability of success (Brown & Niemi, 2007; Ruiz, 2007). Brown and Niemi (2007) and Armstrong (1994) noted that inaccurate referral to remedial courses increases cost for students, extend time-to-degree completion, and misallocate educational resources of the institution.

Ruiz (2007) concluded that, "the increase in the number of developmental courses offered at community colleges may not be an accurate picture of the skill level of [our] students" (p. 190). It is stipulated that many such courses merely serve as review courses, rather than teaching new material (Thompson, 2006; Ruiz, 2007). Students could be remediated via refresher courses, thus accelerating their time-to-degree completion rates and reducing educational expenses (Thompson, 2006; Ruiz, 2007).

Conflicting policy signals. The next section addresses the lack of articulated standards and clear signals concerning adequate academic preparation by community colleges. Kirst (2003) and Kirst and Bracco (2004) remarked that policies concerning the transition from secondary to postsecondary education are fragmented and confusing. However, policy signals and incentives are crucial factors in stimulating students' desire to obtain the knowledge required for postsecondary academic success (Kirst & Bracco, 2004). Students often do not know what colleges expect of them in terms of meeting their admission requirements (Kirst & Bracco, 2004;

Ruiz, 2007). Many believe that nonselective four-year institutions and community colleges do not have academic standards and there is little penalty for poor high school performance (Kirst & Bracco, 2004; Ruiz, 2007). But community colleges have academic standards as evidenced by the widespread use of placement tests for access to credit-level courses (Brown & Niemi, 2007; Kirst & Bracco, 2004; Venezia, 2000). Kirst and Bracco (2004) concluded that, "clear, consistent, and appropriate signals and incentives have shown to improve student learning and affect students' motivation positively" (Kirst & Bracco, 2004, p. 4).

Students and attitude. Research professionals in the education community identified student behavior and attitude as one of the key contributors to the proliferation of postsecondary developmental courses (Hager, 2005; Kirst, 2003; Kirst & Bracco, 2004; Ruiz, 2007; Thompson, 2006). High school students assume that high school graduation requirements are sufficient for college-level coursework (Kirst, 2003; Kirst & Bracco, 2004; Smydo, 2008). They realize that they will be admitted to broad access institutions if they meet minimum GPA and course requirements (Kirst, 2003; Kirst & Bracco, 2004). However, rarely are students warned that they may have difficulty completing a degree if they do not take certain courses and achieve certain grades in high school (Kirst & Bracco, 2004). So students often take no math during their senior year in high school, therefore they are not prepared for placement testing and college-level math coursework (Blair, 2008; Kirst, 2003; Kirst & Bracco, 2004; Thompson, 2006).

A study by Ruiz (2007), investigating intervention strategies to reduce the number of students who may be incorrectly placed into developmental math courses, revealed that even though 85% of students were aware of the placement test requirement, 85% did not study for the test. Consequently, they were not prepared for placement exams and ended up in non-credit remedial courses (Kirst, 2003; Ruiz, 2007). Students in the study further admitted that they did not feel challenged in the mathematics course they enrolled since they were merely reviewing material they had learned in high school (Ruiz, 2007).

Concluding the discussion on possible causes of the remedial math dilemma from the perspective of high schools, community colleges and students, the next section will turn to an examination of the historical split inherent in the K-16 system and ramifications thereof.

K-16 policy mechanism. The content between high school exit exams and college admission and placement tests differs (Brown & Niemi, 2007; Kirst, 2003; Kirst & Bracco, 2004; Venezia, 2000). The difference in content leads to confusion about what students really need to know in order to succeed in college (Kirst & Bracco, 2004; Smydo, 2008). Kirst and Bracco (2004) noted that,

> the causes of remediation, non-completion, and inadequate secondary preparation lie in part in the historical split between levels of our educational system and the subsequent lack of communication and connection between them. Public education in the United States essentially comprises two distinct levels: elementary and secondary (K-12) and postsecondary, or higher, education. (p. 2)

Over the years, there has been very little coordination of reform across educational levels, and college and university reformers as well as K-12 reformers have failed to look at changes in postsecondary education (Kirst, 1998; Kirst & Bracco, 2004; Venezia, 2000). More and more people are calling for a K-16 perspective on education, ideally all one system (Kirst & Bracco, 2004). However, few policy mechanisms address the connection between secondary and postsecondary education and most policies simply perpetuate the separation (Kirst & Bracco, 2004; Venezia, 2000). Another phenomenon still present in postsecondary education is the underlying notion of many K-12 policies that not every student is going to college (Kirst &

Bracco, 2004). But reality reflects that today over 70% continue on to college. As this belief is misguided and obsolete, it may negatively affect many students (Kirst & Bracco, 2004).

Conclusion. The disjuncture between K-12 education and postsecondary education is uniquely American (Kirst, 1998, 2001). There is a much greater disjuncture between secondary and postsecondary education in the U.S. than in most other nations (Kirst, 2001). One suggestion for improving the disjunction between high school and community colleges is to develop continuity across the K-16 system and convey to students early in their educational career the academic expectations of community colleges (Brown & Niemi, 2007; Kirst, 2003; Kirst & Bracco, 2004; Roth et al., 2001; Venezia, 2000). Furthermore, community colleges need to outline course and skill requirements needed for college-level credit-bearing coursework and how they relate to courses taken in high school (Brown & Niemi, 2007; Roth et al., 2001; Venezia, 2000). They also need to ensure that such information is communicated to students, parents, high school counselors, and high school district curriculum personnel on a consistent and ongoing basis (Brown & Niemi, 2007; Roth et al., 2001).

One way of sending a uniform message and generating consistent expectations across community college campuses is to standardize placement practices, i.e. develop a placement test specifically for community colleges (Brown & Niemi, 2007; Kirst, 1997; Venezia, 2000). Such a dramatic shift would require involvement at the state level, with community college boards playing a major role in establishing a consistent and coherent placement testing policy (Brown & Niemi, 2007). Additionally, community college placement tests would need to be aligned with the skills actually addressed in the community college classrooms (Brown & Niemi, 2007; Kirst, 1997; Kirst & Bracco, 2004). Lastly, bringing K-12 education leaders into the discussion would allow a collaborative effort between the State Board of Education and community college boards to ensure consideration of both perspectives in aligning secondary and postsecondary systems (Brown & Niemi, 2007; Kirst, 1997; Kirst & Bracco, 2004; National Center, 2008; Smydo, 2008; Venezia, 2000).

It is time that our nation addresses the disjuncture of our educational system. We need to "make a comparable effort to send clear signals, properly inform, and prepare those students who will enroll at the community college, especially since they constitute the vast majority of college students" in the United States higher education system (Brown & Niemi, 2007, p. 28).

As the discussion on possible causes contributing to the remedial math dilemma has reach conclusion, the next section will examine the various assessment placement practices at community colleges.

Assessment Placement at Community Colleges

This section avails not only a historical account of higher education standards and college preparation in the United States, but also an in-depth examination of current assessment placement processes, policies, and instruments used at community colleges.

History of higher education standards and college preparation. In the United States a pervasive disjuncture exists between lower and higher education, which stems in part from the way the nation created mass education systems for both K-12 and higher education (Kirst, 2001). In Europe universities play a major role in determining the content of secondary curriculum as well as the content and format of secondary school examinations (Kirst, 2001). At one time United States colleges and universities also played an important role in high school curriculum (Kirst, 2001). In 1900, the College Board set uniform standards for academic subjects to help students prepare for college entrance examinations (Kirst, 2001). However, in the postwar years, the notion of K-16 academic standards vanished and aptitude tests like the SAT replaced subject matter standards for college admission (Kirst, 2001). The only nationally aligned K-16 standards efforts, which remains in effect today, is Advanced Placement (AP) (College Board, 2009a; Kirst, 2001; Mitchell, 2006).

Today secondary and postsecondary schools work in separate orbits within each state and at the federal level, and are guided by policies exclusive to each sector (Brown & Niemi, 2007; Kirst, 2001, 2003; Kirst & Bracco, 2004). Mediating professionals employed by high schools and colleges rarely interact (Kirst, 2001). A universal emphasis on access and admission to college rather than preparation, has become the theme of K-16 policy, high school counselors, college recruiters, college admissions and financial aid officers, as well as students and parents (Kirst, 2001).

Frequently high schools are blamed for low standards and inadequate preparation of students for college-level performance (Kirst, 1997). Various studies have investigated the relationship between remedial education in college and students' high school preparation (Abraham & Creech, 2000; Biswas, 2007; Collins, 2008; Golfin et al., 2005; Kirst, 2001; Morgan & Michealides, 2005; Ruiz, 2007). Many believe that students enrolled in college preparatory courses should be less likely to require remedial math, but most results revealed the contrary (Kirst, 2001; Ruiz, 2007). A study by Hoyt and Sorensen in 2001 found that more than half of the students, who completed three years of college preparatory math, were placed in remedial math and almost 25% of students who took calculus in high school tested into Intermediate Algebra (as cited in Ruiz, 2007). A 2007 study by Achieving the Dream Community Colleges Count revealed that remediation rates for students completing their highest level of mathematics in grade 12 were: (a) Algebra II or equivalent-- 63%, (b) Advanced mathematics, such as algebra/trigonometry—32%, (c) the next higher level of mathematics, such as pre-calculus—15%, and (d) the most advanced mathematics course, i.e. calculus—3% (Biswas, 2007). One study that provided supporting evidence was conducted by Lappan and Phillips, concluded that students who have taken three years of college preparatory courses with a B average generally are successful in college-level math courses (as cited in Ruiz, 2007). In lieu of the aforementioned trends, the question remains if the high school curriculum lacks rigor or if there is a problem with the placement test remains (Ruiz, 2007).

Current reforms have attempted to create a seamless transition between secondary and postsecondary education, but have ignored the lack of coherence in content and assessment standards between K-12 and higher education (Biswas, 2007; Brown & Niemi, 2007; Kirst, 2001; Kirst & Bracco, 2004; Prince, 2005). As a result of the K-16 disjuncture, many students face three conflicting standards, namely high school graduation, college admissions and college placement, and they have no clear sense of what knowledge and skills constitute adequate academic preparation for higher education (Brown & Niemi, 2007; Kirst, 2001; Kirst & Bracco, 2004; Prince, 2005).

Colleges are aware of this trend and know how many entering freshmen fail their placement tests or drop out because they are not academically prepared for college-level work (Kirst, 2001). But most colleges, like their applicants, are more concerned about access to higher education than academic preparation needed to complete a postsecondary degree or certificate (Kirst, 2001). This trend is evidenced by the fact that most community colleges have open admission policy and as such provide few incentives for high school seniors to take rigorous academic courses (Brown & Niemi, 2007; Kirst, 2001, 2003; Kirst & Bracco, 2004).

In 2001, the U.S. Department of Education recommended, in an effort to increase the coordination between K-12 schools and colleges and universities, improving college admission and placement practices deserves a high priority (Kirst, 2001). A 2008 Achieving the Dream policy brief stressed that greater emphasis be placed on policies governing student assessment and placement in developmental education (Biswas, 2007; Collins, 2008).

The purpose of this section is to examine current assessment and placement practices at community colleges, investigate which assessment instruments produced the most reliable results, and provide insight into the validity of placement tests.

Assessment and placement practices. Assessment placement processes at community colleges remain one of the most debated aspects of remedial education. Placement processes and the various placement tests and evaluation tools used to assess student proficiencies will be discussed in the following section.

Placement processes at community colleges. Almost every community college has some type of placement instrument to determine college readiness of entering students (Felder, et al., 2007; Ternes, 2009). Placement tests represent the defacto Math and English standards at community colleges (Felder et al., 2007; Pavelcheck, Stern, & Olson, 2002). The placement process at community colleges is critical for several reasons (Felder et al., 2007): (a) Community colleges are open access institutions that serve a heterogeneous student population; (b) students range from high school dropouts to high school graduates to adults returning to pursue a college education; and (c) students are not required to make formal application and many come without high school transcripts. Therefore community colleges need a way to measure students' abilities and sort students into classes (Abraham & Creech, 2000; Felder et al., 2007).

Placement at community colleges is considered a high-stakes test, a test that has consequences attached to the results, as it can greatly impact students' college outcomes (Felder et al., 2007; Mitchell, 2006; Morgan & Michaelides, 2005; Ruiz, 2007). Beginning courses at the right level is critical for students, "given that approximately 50% of students will be gone within one year" (Felder et al., 2007, p. 4; ICHE, 2008; Kirst & Bracco, 2004). Starting college in remedial courses is risky; very few students, approximately 10%, make the transition from developmental to transfer-level courses, or stated differently nearly 90% of students starting in a developmental course, in particular those that start two levels below college math, never pass a transfer-level course (Brown & Niemi, 2007; Felder et al., 2007). A study by the National Bureau of Economic Research, an analysis of nearly 100,000 community college students in Florida, found that those on the edge of needing remediation and being placed in a remedial course has the effect of increasing the odds that students will return for a second year (Calcagno & Long, 2008; Jaschik, 2008). Only a sound placement test will provide reliable results. Ternes (2009) so succinctly remarked,

> Assessment is the key to correct placement, and correct placement is where the success of any developmental program starts. If you build a building with a poor foundation, it doesn't matter what you put on the top, the building is bound to have problems. A wonderful remedial program will not compensate for a weak assessment and placement system. (para. 1)

Placement exams at community colleges are numerous, varied, largely unknown, and designed to serve the assessment needs of departments within postsecondary institutions (Brown & Niemi, 2007; Kirst, 2003; Kirst & Bracco, 2004; Pavelcheck et al., 2002). These exams are extremely diverse with many institutions using academic departmental faculties to devise a local exam while others use commercially developed standardized tests (Brown & Niemi, 2007; Felder et al., 2007; Gerlaugh et al., 2007; Kirst, 2001; Shults, 2000). In addition, such exams are not well published to high school students and their parents or to teachers, nor are they aligned with current K-12 standards (Brown & Niemi, 2007; Kirst, 2001; Kirst & Bracco, 2004; Venezia, 2000). As noted previously, hundreds of variations of placement exams exist (Kirst, 2003), thereby reflecting great variability and inconsistency throughout the community college system (Brown & Niemi, 2007; Kirst & Bracco, 2004).

Since placement tests used by community colleges must accommodate many students, standardized tests are the most economical choice, require little effort to implement, and are easily accessible and maintained (Ruiz, 2007). However, placement tests are a source of controversy as they may result in academic tracking by ability and possibly exclude students rather than aid access (Brown & Niemi, 2007; Kirst, 2001; Kirst & Bracco, 2004; Ruiz, 2007). In order to understand the nature, purpose, and process of assessment and placement at community colleges, an explanation of various testing instruments and assessment practices, their underlying design along with a discussion of validity and reliability is presented. To understand the various options available to community colleges, a description of types of assessment and placement tests and utilization thereof follows.

Types of assessments and placement tests. As placement tests take on many formats, a discussion of the types of assessment tests currently used by community colleges is warranted. A wide variety of college readiness tests exist and while they take on many forms, they also share many of the same features including: (a) Most include a math test (typically with multiple levels), (b) are computerized and adaptive, (c) are standardized, and (d) are considered high-stakes tests (Ternes, 2009). Placement tests in mathematics at community colleges generally assess students' skills in algebra, geometry, trigonometry, and pre-calculus (Kirst, 2003). A few commercially distributed tests dominate the field, such as COMPASS and ACCUPLACER, other tests in widespread use are MDTP, ASSET, and college placement tests locally developed by math faculty on an institutional basis (Brown & Niemi, 2007; Felder et al., 2007; Gerlaugh et al., 2007; Golfin et al., 2005; Kirst, 2001; Kirst & Bracco, 2004; Ruiz, 2007; Ternes, 2009). Supplementary tests used for placement and assessment include ACT, SAT, AP, CLEP, TABE, and high school exit exams (College Board, 2009a; Felder et al., 2007; Golfin et al., 2005; Kirst, 2001; Mitchell, 2006). Prior to distinguishing between various tests an explanation of types of assessments is warranted.

Formative vs summative assessment. Over the years, two kinds of assessments have evolved—formative and summative. Formative assessment is used to measure the academic progress students have made and determine if and what adjustments are needed (Mitchell, 2006). Most formative assessments are teacher-designed, although textbook publishers have begun supplying assessment tools, e.g. tests on CD-Rom accompanying the textbook (Mitchell, 2006). Summative Assessment records the state of student learning at a certain point in a students' academic career—at the end of a school year, high school graduation, college entrance, etc. (Mitchell, 2006). Currently two types of summative assessment exist, namely norm-referenced and criterion-referenced standardized tests. These types of assessments are used for high-stakes testing by states and community colleges, and as such have generated much controversy (Mitchell, 2006). A definition of standardized and high-stakes tests as well as norm-referenced and criterion-referenced tests are provided below.

Standardized testing. Standardized testing assumes that certain conditions are met: (a) One test is administered simultaneously to all students under identical conditions, and (b) scoring of the test is predetermined (Mitchell, 2006). If the aforementioned conditions are satisfied the results can be used to compare student performance across schools, districts, states or colleges, and eliminate or minimize variations in results due to difference in administration or form of the test (Mitchell, 2006). Standardized tests can have almost any format, not just machine scored multiple-choice tests (Mitchell, 2006).

Low- and high-stakes testing. Low-stakes testing, a form of formative assessment, has no consequences outside the school or college classroom. Examples of low-stakes assessment tools are weekly quizzes, test, or writing assignments, and the result generally contributes to a student's grade (Mitchell, 2006). On the other hand, high-stakes tests have consequences outside the school or college classroom. High-stakes testing can be used to determine students' promotion from grade to grade, graduation from high school, or admission to college. Results of college readiness testing can have a significant impact on a student's success in college-level courses and college completion rate (Felder et al., 2007; Kirst, 2001; Ruiz, 2007). Placement tests at community colleges are high-stakes tests that can impact students' college outcomes (Felder et al., 2007). Classification of summative standardized, high-stakes testing can be further expanded to distinguish between norm-referenced and criterion-referenced testing.

Norm-referenced testing. Norm-referenced tests do not measure students' achievement based on academic standards, but compare students' achievement to other students, i.e. a norm group (Mitchell, 2006). The design is governed by the normal or bell-shaped curve. Norm-referenced tests spread students' scores along the bell curve and results are frequently reported in percentiles with a score at the 50th percentile the average (Mitchell, 2006). Percentile rankings are often used to identify students for various academic programs, including remedial courses (Mitchell, 2006).

In addition, questions asked in the test do not necessarily represent the knowledge and skills that all students are expected to have mastered (Mitchell, 2006). Test items answered correctly by 80% or more of the test takers are removed from the test during the pilot phase (Mitchell, 2006; Ruiz, 2007). Thus, norm-referenced tests have led to frustration for teachers and students (Mitchell, 2006). From a teacher perspective, teaching important and relevant skills and knowledge may not produce a high success rate for students taking the test. For students, the test causes frustration since no group of students can achieve higher levels without others achieving at lower levels (Mitchell, 2006). The test makes it mathematically impossible for all students to score above average (Mitchell, 2006). Hence, it is imperative for testing administrators to ensure that the chosen norm is a useful comparison for their students and that such a norm has been developed recently (Mitchell, 2006; Ruiz, 2007).

Criterion-referenced tests. Criterion-referenced tests, also called standard-based tests, are designed to compare students' achievement to a given standard (Mitchell, 2006). Contrary to norm-referenced tests, it is possible for all students to achieve the highest or the lowest score since the test does not compare students and does not assume an underlying normal distribution of test scores (Mitchell, 2006). Results are reported in levels, such as basic, proficient, and advanced. Test items are chosen to measure mastery of essential knowledge and skills based on a set of standards (Mitchell, 2006).

Reliability, validity, and fairness. All standardized tests must meet psychometric standards for reliability, validity, and lack of bias (Mitchell, 2006). Reliability infers internal consistency of a test, i.e. a student taking the test repeatedly will achieve approximately the same score (Mitchell, 2006). Validity implies accurate measurement of what the test is intended to measure (Mitchell, 2006). Unbiased testing implies fairness, which is the key requisite for any reliable and valid test (Mitchell, 2006). Students must not be at a disadvantage with respect to ethnicity or socioeconomic status (Mitchell, 2006).

Colleges are accountable to students, staff and the taxpayer, and assessment is the essential ingredient of accountability (Armstrong, 1999; Mitchell, 2006). But are institutions of higher education accountable with respect to assessing academic progress of students and reliability of tests using current testing instruments and procedures? Ternes (2009) remarked that, "almost all of the tests currently in use are poorly understood and under researched" (para. 1). Research on the efficacy of placement testing is mixed (Armstrong, 1999; Brown & Niemi, 2005). In 1999, Armstrong found that studies relating community college placement test scores to course outcomes resulted in a small or modest relationship with low to moderate validity coefficients. In 2005, Goldberg pointed out that egregious examples of errors are prevalent during the administration and scoring of tests (as cited in Mitchell, 2006). Although such accusations were quickly dismissed claiming that errors affect only a small percentage of schools and students due to the huge amount of tests administered (Mitchell, 2006). Most tests are commercially developed and distributed, and the testing industry is responding to the enormous demand for large-scale testing with quick and easy testing solutions (Mitchell, 2006). In 2003, the vice president of Princeton Review remarked that, the testing industry is "still in the primitive … knives and bearskins stage," and pressure for more and better testing will increase (as cited in Mitchell, 2006, p. 4).

Application of criterion-referenced testing. This section discusses application of criterion-referenced tests at the K-12 and college level.

K-16 statewide assessment and high school exit exams. Since the 2001 NCLB implementation the use of the number of criterion-referenced tests has dramatically increased with 44 states using such tests to report Adequate Yearly Progress (AYP) to the U.S. Department of Education as part of K-12 assessment (Mitchell, 2006). At the college-level, such a movement has been missing as most colleges use norm-referenced tests with the exception of placement tests developed on a statewide or institutional basis (Mitchell, 2006).

Some states have established statewide standards and require students to take a statewide exam in their junior or senior year in high school. These tests generally serve to determine how well students meet academic standards, and in some cases, students who pass the exam are exempt from taking remedial education courses at state colleges (NCPPHE, 2008). For example, in California high school students take the augmented California Standards Test (CST), an Early Assessment Placement (EAP), developed by the California Department of Education, the State Board of Education, and the California State University (CSU) System to determine if high school students are ready to enroll in college-level courses at CSU, and students who pass the exam are exempt from developmental education courses (NCPPHE, 2008). Massachusetts and Texas also administer high school exit exams, namely the Massachusetts Comprehensive Assessment System (MCAS) and Texas Assessment of Academic Skills (TAAS), respectively. Other high school tests administered are the Stanford 9, published by Harcourt Brace Educational Measurement, TerraNova, published by CTB/McGraw-Hill, New York State Regents, Kentucky Commonwealth Accountability Testing System (CATS), etc. (Education Trust, 1999).

NAEP. The National Assessment of Educational Progress (NAEP), also known as the Nations Report Card, uses criterion-referenced tests (Mitchell, 2006). NAEP tests students in 4th, 8th, and 12th grade every two years in reading and math and at longer intervals in other subject areas (Mitchell, 2006). Test items are derived from the NAEP Frameworks, which lists the national standards for NAEP (Mitchell, 2006). The NAEP has an inherent potential for aligning nationwide measures of college readiness (Kirst, 2003).

CLEP. The College-Level Examination Program (CLEP) offers students the opportunity to earn college credit for most courses taken in the first two years of college (College Board, 2009c). CLEP exams, primarily made up of multiple-choice questions, are developed by faculty members at institutions across the country and are criterion-based (College Board, 2009c). Most colleges grant credit to students earning satisfying scores on such exams (College Board, 2009c). The criterion-based feature ensures that the score earned by one student is not dependent upon how well other students scored (College Board, 2009c). CLEP scores are not comparable to SAT scores or other test scores (College Board, 2009c).

TABE. The Test of Adult Basic Education (TABE), produced by CTB McGraw-Hill, is a less commonly used test developed by content specialists to determine common educational goals, knowledge, and skills emphasized in the curricula (Golfin et al., 2005). The test encompasses five graduated levels of difficulty across five different content areas including reading, language, and applied math (Golfin et al., 2005). The test is available in both paper-and-pencil and software format (Golfin et al., 2005).

MDTP. Math Diagnostic Testing Project (MDTP) is a placement test developed by faculty from the University of California (UC) and CSU prior to the existence of COMPASS (Felder et al., 2007) and widely used in the state of California (Brown & Niemi, 2007; Felder et al., 2007; NCPPHE, 2008).

AP Exams. The College Board developed the Advanced Placement (AP) program to provide assistance in the transition to higher education and the awarding of college credit (College Board, 2009a; Kirst, 2003; Mitchell, 2006). The AP program is the only cooperative educational endeavor between secondary schools and higher education allowing students to get credit for college level work while in high school (College Board, 2009a; Kirst, 2001; Mitchell,

2006). While some colleges will award college credit to students with a score of three or higher, policies on the acceptance of AP scores varies among different institutions (Brown & Niemi, 2007; Kirst, 2001; Mitchell, 2006). AP is mostly used by selective four-year institutions (Kirst, 2003).

Application of norm-referenced standardized testing. Most commercially developed standardized tests are norm-referenced test including SAT, ACT, ACCUPLACER, COMPASS, and ASSET. A discussion of such tests is provided below.

SAT and ACT. Many colleges and universities use SAT or ACT math scores to place students in appropriate math courses (Mitchell, 2006; Foley-Peres & Poirier, 2008). Virtually all four-year colleges and universities require results from at least one of these tests for admission (Golfin et al., 2005; Foley-Peres & Poirier, 2008). Since admission is open to everyone community colleges do not require the ACT or SAT (Golfin et al., 2005). Some community colleges will accept scores from either one of these tests in lieu of the institutionally required test to determine students' eligibility for college-level math (Golfin et al., 2005). However, many community colleges require students to take an assessment test despite the availability of ACT or SAT scores to ensure correct course placement (Golfin et al., 2005). A few states, including Illinois and Colorado, mandate that every student in grade 11 must take the college admission test used in their states (Golfin et al., 2005; Mitchell, 2006).

The SAT, owned by the College Board but designed and administered by the Educational Testing Service, was created to predict college success and deliberately disconnected to any state or school curriculum (College Board, 2009a; Mitchell, 2006). The scores would provide evidence of student credentials and eliminate the need to sit for entrance exams at various institutions. Its secondary purpose was to eliminate the bias of inconsistent grading systems and curriculum standards throughout the nation's high schools (College Board, 2009a). The test, regarded by many as an aptitude test rather than an achievement test, was recently revamped to better reflect the high school curriculum (Mitchell, 2006; Choroszy & Muehlberg, 2006).

ACT, was created by ACT and is designed to measure mastery of skills and knowledge essential for success in college freshman courses (ACT, 2009; Mitchell, 2006). The assessment test scores are typically used for college course placement as they are based on the generalized curriculum described in the Standards for Transition or College Readiness Standards (Mitchell, 2006).

ACCUPLACER, COMPASS, and ASSET. The most commonly used assessments tests at community colleges are ACCUPLACER, COMPASS and ASSET, all of which are commercial standardized norm-referenced tests. Although each assesses skill levels in three academic areas, namely reading, writing and math, this research study focuses only on the math portion of ACCUPLACER, COMPASS and ASSET (Ruiz, 2007). As noted earlier, norm-referenced tests do not measure student achievement based on academic standards (Mitchell, 2006). Commercial standardized norm-referenced tests pass through many revisions to ensure validity, and in the pilot phase questions that are mostly answered correctly or incorrectly are eliminated from the test (Mitchell, 2006; Ruiz, 2007). Therefore, as Ruiz (2007) noted, it is unclear whether the remaining questions are truly representative of the knowledge the instrument is trying to measure.

ACCUPLACER and COMPASS are both un-timed computerized, adaptive instruments, while ASSET is a written timed 25-minute test (Ruiz, 2007). Each test incorporates various test levels with varying difficulty levels: (a) ASSET has four testing levels: Numerical Skills, Beginning Algebra, Intermediate Algebra, and College Algebra; (b) COMPASS has three testing levels: Pre-Algebra, Algebra, and College Algebra; and (c) ACCUPLACER also has three testing levels: Arithmetic, Algebra, and College Level Math (Ruiz, 2007).

Each of these placement testing instruments requires students to select a particular test level based on academic experience or comfort level, generally assisted by a recruiter or advisor (Ruiz, 2007). Ruiz (2007) observed that, "selecting a test level may be overwhelming to students and may have a [significant] impact on the course placement," i.e. college-level versus remedial (p. 57). However, most students are not aware of the consequences selecting a test level may have on their placement (Ruiz, 2007). Ruiz's study showed that all students who were not given information regarding the test levels were placed in a developmental math class. Therefore, selection of a particular testing level can result in a different placement level, and students who select the lowest test level will automatically limit their enrollment options to developmental math (Ruiz, 2007). For example, the design of the Numerical Skills test of ASSET is to assess basic arithmetic knowledge and the highest level course a student would be able to enroll is Beginning Algebra, a developmental math course at most community colleges (Ruiz, 2007). Conversely, a student who selects the Elementary Algebra level of ASSET may have the opportunity to enroll in a college-level math course (Ruiz, 2007). As a result, students who inadvertently select the Numerical Skills test even though they had taken higher levels of math in high school, will be incorrectly placed in developmental math courses most likely several levels below college math (Ruiz, 2007).

ACCUPLACER, was developed by the College Board and on the market since 1986, is a computer-adaptive placement test that quickly assesses reading, writing and math skills, and on average students are asked to answer on average 17 questions (Brown & Niemi, 2007; College Board, 2009b). According to a study by the U.S. Department of Education, ACCUPLACER is the third most commonly used placement test by two-year postsecondary institutions. In addition, some high schools use this test in conjunction with local community colleges (Golfin et al., 2005).

The test produces score reports containing student test scores and information about course placement based on institutional placement scores (College Board, 2009b). The results of the assessment, in conjunction with student academic background, goals, and interests, can be used by academic advisors and counselors to determine course selection or placement (College Board, 2009b).

ACCUPLACER is an adaptive test, i.e. questions are chosen based upon responses to previous questions (Mekonnen & Reznichenko, 2007). The test supposedly selects just the right question dependent on a student's ability level. The first question in a particular test is of average difficulty chosen randomly from a set of starter questions of the same level of difficulty (Mekonnen & Reznichenko, 2007). Hence, no two tests taken are precisely the same and raise the question as to the validity of this standardized test (Mekonnen & Reznichenko, 2007). The student can change the answer to a particular question before moving on to the next question, but cannot omit a question or come back to the question at a later time to change the answer (Mekonnen & Reznichenko, 2007). Regulations regarding calculators, textbooks, notebooks, etc, are set by the institution (College Board, 2009b). The math portion tests students on the levels of proficiency in each of the three categories on a percentage basis (see Appendix A): (a) Arithmetic—16%, (b) Elementary Algebra—63%; and (c) College Level Math—21% (College Board, 2009b; Golfin et al., 2005). Please note that Arithmetic, Elementary Algebra I and College Level Math represented in the table are Algebra, Algebra II, and Trig/Precalculus, respectively. The College Board provides suggested cutoff scores, but determination of such is left to the institution (Brown & Niemi, 2007; Golfin et al., 2005). For test preparation the College Board suggests SAT preparation or SAT review books (College Board, 2009b).

Even though COMPASS, developed by ACT, requires students to only answer 7-14 questions, it is the most frequently used test at two-year higher education institutions. Usage of ASSET is a distant second, and some institutions use both (Brown & Niemi, 2007; Golfin et al., 2005). COMPASS and ASSET tests cover much the same material, except that the COMPASS test is an un-timed, computerized adaptive test, while ASSET is a timed paper-and-pencil test (Golfin et al., 2005; Ruiz, 2007). ASSET has four testing levels: Numerical Skills, Beginning Algebra, Intermediate Algebra, and College Algebra; and COMPASS has three testing levels: Pre-Algebra, Algebra, and College Algebra (Brown & Niemi, 2007; Ruiz, 2007). Questions on COMPASS (see Appendix A) are devoted to the following topics on a percentage basis: Algebra 1—14%, Geometry—23%, Number Theory, Arithmetic, Combinatorics, Logic—19%, Algebra 2—25%, and Trig/Precalculus—15% (Kirst, 2001). The adaptive feature of the test adjusts to each student's ability, and the difficulty level of questions varies depending on how many questions the student answers correctly (Felder et al., 2007). Thus, no two students are given the same set of questions (Felder et al., 2007), and as such questions the standardization requirement of the test. Standardized norm-referenced tests also raise questions about fairness as poor and minority students score lower than White students. As Ruiz (2007) remarked there is more to the correct placement of students than their test scores.

Content comparison of standardized tests. According to a content analysis by RAND significant differences exist in the content between assessment tests used in secondary school compared to postsecondary (Kirst & Bracco, 2004) and as such will be discussed in detail below.

ACT and SAT vs ACCUPLACER, COMPASS, and ASSET. The differences in content between ACT, SAT, and high school math assessment and college placement tests are enormous (Education Trust, 1999; Brown & Niemi, 2007; Kirst, 2001; Kirst & Bracco, 2004). ACCUPLACER, COMPASS, and ASSET emphasize intermediate algebra skills (Algebra II) and trigonometry/precalculus skills (Gerlaugh et al., 2007; Kirst, 2001). The content of MDTP and CPT's developed statewide by a consortium of colleges or locally developed by math faculty on an institutional basis varies, but in general focuses on topics similar to those of ACCUPLACER, COMPASS, and ASSET (Brown & Niemi, 2007; Gerlaugh et al., 2007, Kirst & Bracco, 2004).

SAT, ACT, and high school math assessment tests tend to focus on basic algebra (Algebra I), geometry, probability and statistics, and numbers theory, but ignore intermediate algebra (Algebra II) and trigonometry/pre-calculus (Kirst, 2001). Thus, high school students prepare for and are admitted to college based on one set of skills, but are faced with passing a placement test that covers different topics (Kirst, 2001). In 2008, almost 57% of ACT-tested students who went directly from high school to community college were not ready for college-level math courses (Blair, 2008; Thompson, 2006). As a result students may be unfairly placed in developmental education and as such raises the question as to the reliability of the tests (Brown & Niemi, 2007; Ruiz, 2007).

High School Math Assessment vs ACT and SAT. An analysis by Education Trust in 2000 showed that in the state of California the content of statewide high school math assessment tests and the content of the math portion of the SAT and ACT are fairly similar (Kirst, 2001). These findings are consistent with a later study conducted by ACT and Education in 2004, which proved that ACT scores positively correlated with success in freshman college courses (Mitchell,

2006). Kirst (2001) remarked that SAT I and ACT provide some national assessment uniformity, but neither were well aligned with recent K-12 reforms in California and advocated adjustments to accommodate the new K-12 standards.

Practices and policies in community colleges. Community colleges across the nation greatly differ in terms of placement practices and policies. These differences are subject to discussion in the pages that follow.

Setting assessment placement policies. While community colleges have open admission policies nearly all require students to take a placement prior to enrollment in credit-bearing college-level courses (Kirst, 2001; Kirst & Bracco, 2004; Prince, 2005). Some colleges and universities allow students to take placement exams in high school (Kirst, 2001; NCPPHE, 2008). According to a 2001 study by American Association of Community Colleges (AACC), the majority (58%) of institutions mandated assessment of basic skills for all students, and 75% of these institutions required placement based on testing (Golfin et al., 2005; Shults, 2000). Of those institutions that required placement, almost two-thirds reported that policy was set by the institution, while the remaining third indicated that policy was mandated by the state (Golfin et al., 2005; Shults, 2000).

A 2008 study by Achieving the Dream reported a growing trend in statewide mandatory placement policies at community colleges (Boroch et al., 2010; Collins, 2008; Prince, 2005). Of the 46 states that participated in the study, today 27 states now have such policies compared to 21 states in 2002 (Collins, 2008). In addition, between 2002 and 2008, the number of states requiring colleges to use one or more approved placement exam almost doubled with COMPASS being the most frequently required exam followed by ACCUPLACER and ASSET (Collins, 2008). Furthermore, the number of states requiring standardized cutoff scores for placement into remedial education virtually quadrupled, from 5 states in 2002 to 19 states in 2008 (Collins, 2008). Many of the remaining states realized that student outcomes could be greatly improved by increasing the accuracy of student placement, but are torn between systematizing their placement policies and protecting institutional flexibility (Collins, 2008).

Criteria for exemptions. Many institutions allow exemptions from required participation in assessment testing and vary greatly among institutions (Golfin et al., 2005; NCPPHE, 2008; Shults, 2000). The most frequently reported exemption was college entrance exam scores (76%) (Golfin et al., 2005; Shults, 2000). Other criteria for waiver of the test included high school grade point average (GPA)—7%, statewide high school exam scores—12%, advanced placement (AP) scores—14%, transfer from another postsecondary institution—58%, and other criteria—27% (Golfin et al., 2005; Shults, 2000).

A 2008 nationwide survey of community college math educators indicated that 34% of institutions used the placement exam as the only criteria for assessment and placement (Knowlton, 2008). The remaining institutions used other criteria in conjunction with the placement exam, with ACT scores, high school grades or both the most commonly used criteria as shown in Table 1 (Knowlton, 2008).

Table 1

Additional Placement Criteria Used by Community Colleges

Placement Criteria	Percentage
None	34
ACT Scores	21
High School Grades	11
ACT Scores & High School Grades	10
SAT Scores	6
ACT & SAT Scores	4
High School Grades, ACT & SAT Scores	2

Placement methods. Institutions used a variety of methods to assess students for placement/advising purposes (Golfin et al., 2005; NCPPHE, 2008; Shults, 2000). The most commonly reported methods were computerized assessment measures (63%) and paper-and-pencil assessment measures (60%). Additional methods reported were college entrance exams (36%), institutionally developed measures (24%), and state-developed measures (16%) (Golfin et al., 2005; Shults, 2000). Cutoff scores on assessment measures were set by the institution in 77% of the cases, and by the state in 23% (Golfin et al., 2005; Shults, 2000).

Location of remediation. Institutions use a variety of organizational administrative structures to provide developmental education courses (Golfin et al., 2005; Shults, 2000). Community colleges offer remedial courses in one of three ways: (a) By subject area, (b) through a separate department, or (c) through one academic department, such as Mathematics (Golfin et al., 2005; Shults, 2000). The majority of institutions (61%) offered remedial courses within their respective subject areas, while 25% housed developmental courses in a separate developmental department, and the remaining 13% offered such courses through one academic department (Gerlaugh et al., 2007; Golfin et al., 2005; Shults, 2000).

Levels of remediation. There is a wide variety of needs in remedial education from students who need a "refresher" course to those who may require multiple courses to achieve college-preparedness (Shults, 2000, p. 6). Developmental math is offered by 97% of public two-year institutions, and 60% of those institutions provide between 2 and 4 courses with an average of 2.5 courses (Golfin et al., 2005). However, the number of levels offered across institutions ranged from one level in various subjects to 16 levels of math (Shults, 2000). Institutions in larger cities tended to have more levels of remediation and enrollment in developmental education in urban institutions was higher than in colleges located in nonurban areas (Golfin et al., 2005; Shults, 2000). Of the institutions offering remedial education, half offered more than three levels of remedial math (Shults, 2000).

Exiting remedial coursework. The majority of institutions (56%) used more than one measure to assess whether a student was prepared for college-level coursework (Golfin et al., 2005; Shults, 2000). Ninety-one percent of institutions used successful completion of remedial courses for assessment (Golfin et al., 2005; Shults, 2000). Other methods of assessing readiness for credit-level courses were exit exams, evaluations, retaking of the original assessment test, and other methods (Shults, 2000).

Institutional limits. In recent years, institutions have received increased pressure to limit the number of times a student may enroll in remedial courses or the number of times the student may enroll in one course (Shults, 2000). In addition, a limit exists with respect to the amount of federal aid a student can receive for remedial education (Shults, 2000). Twenty-six percent of all institutions limit the amount of time a student may spend in developmental courses (Golfin et al.,

2005). The institutions set Seventy-one percent of those limits, while 24% were set by the state (Golfin et al., 2005). Limits with respect to the number of retakes per course, semester or year also exist, and in 86% of institutions such policy was institutionally imposed (Shults, 2000). Restrictions with respect to taking college-level courses not related to a degree or certificate program while in developmental courses virtually does not exist (Golfin et al., 2005).

Test result controversy. Mekonnen and Reznichenko (2007) and Ruiz (2007) stressed that placing students in the appropriate level is a difficult task. Institutional placement processes have a profound influence on access to college-level course, and it appears that current placement practices possibly inflate the number of students placed in remedial math (Ruiz, 2007). Aspects that possibly contribute to misplacement of students and consequences thereof are discussed in this section.

Misclassification of students and cutoff scores. "While the value of placement tests is known, research has shown that not all assessment tests accurately place students in developmental math courses" (Golfin et al., 2005, p. 25). Placing students in the appropriate level is a difficult task particularly when it comes to cutoff scores (Mekonnen & Reznichenko, 2007; Ruiz, 2007). Most colleges face the challenge to determine optimal cutoff scores, but an optimal cutoff score proves to be a rather elusive value (Mekonnen & Reznichenko, 2007; Ternes, 2009). Thus, most institutions rely on cutoff scores provided by the publisher of standardized tests (Mekonnen & Reznichenko, 2007). After all commercial standardized tests, such as ACCUPLACER, are expected to have a 90% reliability (Mekonnen & Reznichenko, 2007).

Golfin et al. (2005) surveyed institutions in various states regarding policies governing the placement of incoming freshmen and focused on the types of tests and cutoff scores used by colleges across the nation. The study revealed that the minimum scores suggested by test publishers for ACT, SAT, ACCUPLACER, COMPASS, and ASSET were higher than those mandated by the states included in the survey (Golfin et al., 2005). As a result the number of students placed in remedial math most likely is inflated.

ACCUPLACER, COMPASS and ASSET cutoff scores. A study by Ruiz (2007) supported the above-mentioned assumption as placement tests, such as ASSET, COMPASS and ACCUPLACER, produced an unquantifiable number of false-negative placements. A false-negative event occurs when a student who scored below the cutscore and considered underprepared for the course successfully completes the course (Ruiz, 2007). False-negative placements relegate students to a course at a lower academic level regardless of true ability, deny students access to college-level courses, and pose a serious issue (Ruiz, 2007). Thus, institutional placement processes have a profound influence on access to college-level courses (Ruiz, 2007).

The study also examined the effect of relaxed cutoff scores and resulted in an increase in the number of students placed in college-level math (Ruiz, 2007). Statistical analysis of mean grades proved that students previously ineligible for college-level courses were 70 to 80 percent as successful as those students eligible (Armstrong, 1999; Brown & Niemi, 2007; Ruiz, 2007). Furthermore, students who enrolled in higher-than-placed mathematics courses within the remedial math sequence were as successful as those placed into the course (Ruiz, 2007).

A paper by Mekonnen and Reznichenko (2007) analyzed the reliability of test scores and evaluated the validity of ACCUPLACER implemented at Baltimore City Community Colleges (BCCC). Despite an effort by the math department at BCCC to revise and reset cutoff scores, the final cutoff scores were only slightly lower for some course levels and identical to the scores suggested by ACT for lower level remedial courses (Mekonnen & Reznichenko, 2007). The study concluded that the ACCUPLACER score did not reflect placement of students in the proper course level (Mekonnen & Reznichenko, 2007). Furthermore, results of the study showed that Elementary Algebra and Arithmetic tests assessed students' skills less consistently than Elementary Algebra and College Math (Mekonnen & Reznichenko, 2007). The study also concluded that the number of questions given for each section was insufficient to properly cover the broad content area of each course (Mekonnen & Reznichenko, 2007). On average the ACCUPLACER required a student to complete a total of 17 questions from Arithmetic, Elementary Algebra, and College Math (Mekonnen & Reznichenko, 2007). These findings negatively impact the validity and reliability of the test (Mekonnen & Reznichenko, 2007).

A 2005 study by Hager at a community college with a mandatory test but not a mandatory placement policy is another example of a high number of false-negative events. The Hager study surveyed students who had recently completed the placement test (COMPASS) and appealed their low placement for a higher-level math course (Ruiz, 2007). The survey indicated that of the 27 students who appealed for a higher course 22 were successful at the higher level, a success rate of 81.5% (Ruiz, 2007). In other words, 22 students were misclassified as underprepared. Such erroneous placement could have been detrimental to a student's college career in addition to requiring extra time and money (Ruiz, 2007).

Accurate assessment of placement test cutoff scores is a way to immediately impact access opportunities for students as such scores prove to be a pivotal point in the placement process and determine into what class a student will be placed (Ruiz, 2007). Movement of the cutoff score in one direction or another impacts the number of students who may be categorized as false negative or true positive (Ruiz, 2007). A 2004 study by Leyba (as cited in Ruiz, 2007) revealed that cutoff scores used for ASSET and COMPASS severely impeded students' enrollment in college-level mathematics. Results of this report suggested alignment of cutoff scores with national norm-referenced scores in the ASSET Reference Manual (Ruiz, 2007).

ACT and SAT cutoff scores. With respect to ACT scores, most community colleges require a score of at least 23 (the suggested cutoff score set by the publisher) on the math component to be eligible to enroll in college-level math courses, but no two-year college uses ACT exclusively to assess basic skills (Golfin et al., 2005). According to a 2003 study by ACT, 66% of all students who took the ACT in 2003, scored 22 or below on the math component and 44% scored below 20 (Golfin et al., 2005). Consequently, several states (AR, CO, FL, OK, SD, TN, and WV) relaxed cutoff scores and used a score of 19 or 20 as the score necessary for students to enroll in college-level math (Golfin et al., 2005). A 2001 study by Hoyt and Sorensen at Utah Valley State College (UVSC) revealed that the average ACT score of entering high school student who took Algebra II and Geometry in high school was 19 or 20 (as cited in Golfin et al., 2005). During this time period, the national average for students who had completed Algebra II was 18 according to a report by ACT, and UVSC required students to score 24 or higher on the ACT math component to be eligible for college-level math (Golfin et al., 2005).

An SAT score of 450 is estimated to be comparable to a score of 19 on the ACT (Abraham & Creech, 2000). In 2000, 46% of students nationwide who took the SAT scored below 450 on the mathematics test (Abraham & Creech, 2000). In Southern Regional Education Board (SREB) states most seniors take the SAT in high school and the percentage of seniors who score below 450 on the mathematics test ranged from 47% to 59% (Abraham & Creech, 2000). Cutoff scores at community colleges in SREB states range from 400 to 480 (Abraham & Creech,

2000). A 2005 study by the U.S. Department of Education showed similar results, namely a range of SAT scores from 400 to 460 in most states across the nation (Golfin et al., 2005).

Since mid-1980s, the SREB has recommended statewide or system-wide standards for student performance and placement of students into college courses. Adoption of such standards would allow educational leaders to more accurately assess academic preparation, performance, and needs of students entering college (Abraham & Creech, 2000). SREB considered the establishment of state policies on college-level assessment placement of utmost importance (Abraham & Creech, 2000).

In summary, collegiate institutions have an ethical responsibility to keep the number of misplaced students at an absolute minimum (Ruiz, 2007). The goal of every institution should be to set cutoff scores that will minimize false-negative results (Ruiz, 2007). Ensuring that the testing instrument is appropriate for the student population and the intended purpose is an integral part of the validity for any cutoff score process (Morgan & Michaelides, 2005). Morgan and Michaelides (2005) recommended that cutoff scores be monitored on a regular basis, at least every five to seven years or earlier if performance issues suggest otherwise, to assure functionality. If it appears that the number of students being placed into a specific course has drastically changed without a sufficient explanation, then it would be wise to determine if current cutoff scores are still valid and adjustment deems warranted (Morgan & Michaelides, 2005).

Consequences of placing students below ability level. It is important to understand the design of a placement instrument and the cutoff scores set by either the publisher or the college (Ternes, 2009). Setting the placement score too low may set students up for failure as students may be placed in courses for which they are underprepared (Collins, 2008; Ternes, 2009). Setting the cutoff scores too high will not only limit access and increase the number of courses students need to take, but also increase time-to-degree completion and cost for students, parents, institutions, and taxpayers (Collins, 2008; Ternes, 2009).

One may think it is better to "err on the side of caution" and set cutoff scores high enough so that students are not set up for failure (Ternes, 2009). Studies have shown that of students who placed in the lowest level of remedial math, i.e. in need of three levels of remedial math, 18% attempted a college-level course with only 14% of those students completing the course (Biswas, 2007). In the end, the success rate for students requiring three levels of remediation was less than 2% (Biswas, 2007).

Hence, weighing the pros and cons of the cutoff score is crucial to mitigate those opposing goals (Ternes, 2009). A statistical analysis by Ternes (2009) showed that pass rates for students placed in the lowest remedial course peaked slightly above the cutoff score and then declined substantially. Ternes articulated that the decline in performance may be due to student attributes, e.g. students with scores slightly below the cutoff score may be under-challenged by the course, resulting in decreased performance.

Students' lack of challenge may explain the phenomenon that occurred at the Roswell branch of Eastern New Mexico University, i.e. a dramatic decline in the pass rate and completion rate between the years 1989 and 1990 (Armstrong, 1991). In 1990, the college started implementing a new placement test, ASSET, in lieu of the placement test developed by the college (Armstrong, 1991). The switch to the new testing instrument was prompted by the advanced features the new test provided, e.g. computerized reporting and sharing information with other schools at a minimal expense (Armstrong, 1991). The intent of the college was to increase student outcomes and accountability to students, staff, and taxpayers (Armstrong, 1991). However, in 1990-91, implementation of the new placement test resulted in a significant decline in pass rate for the math sector from 46% to 34% compared to 1989-90 (Armstrong, 1991). Completion rates of students in developmental math classes followed a similar pattern, with 96% in 1989-90 and 88% in 1990-91 (Armstrong, 1991).

Another study by Achieving the Dreams produced similar results (Biswas, 2007). At Housatonic Community College (HCC) in Bridgeport, CT, one of the Achieving the Dreams Colleges, success rates for students in remedial math declined from 50% to 39% following the adoption of ACCUPLACER assessment test (Biswas, 2007). Biswas (2007) remarked that,

> the college [lost] 50% of the students in each developmental math level, before students even [reached] the first college-level math course. The college discovered that one of the main reasons for the high dropout rate was the frustration and expense of repeating a course from scratch even when students had mastered some, if not all, of the course content. (p. 3)

Improving testing and placement. Various efforts undertaken by community colleges to improve assessment placement of students have had a positive effect on remediation rates. The most commonly used strategies are discussed below.

Use of multiple measures. Most community colleges place students solely on their placement score (Ruiz, 2007). Researchers have found that students are better placed when multiple measures are incorporated in the advising and placement process (Armstrong, 2000; Boroch et al., 2010; Marwick, 2004; Mekonnen & Rezninchenko, 2007; Ruiz, 2007). Armstrong (2000) studied the predictive validity of the placement test on course grades as well as the effect of student characteristics on course grades, in particular demographic and dispositional variables. The study found that the best predictor of course grade was a combination of the test score and dispositional student characteristic, specifically high school GPA, grade in last math course, and number of years of mathematics in high school (Armstrong, 2000; Ruiz, 2007).

A study by ACT concluded that using the ACT assessment score and high school subject area average jointly improved the accuracy of placement decisions versus using a single predictor (ACT, 1993). The number of high school courses taken in a particular subject area proved to be the least effective variable for making placement decisions for college-level courses (ACT, 1993). Ruiz (2007) substantiated the above findings noting that students who had 3 years of college preparatory math achieved a score of 20 on the ACT, and students with 4 years of college preparatory courses did not reach a cut-score of 24 either.

A study by Foley-Peres and Poirier (2008) compared the use of SAT scores and/or institutionally developed math placement scores for placement of freshman students. The initial study used student SAT scores, college math assessment scores, midterm grades, and college faculty assessments of students enrolled in math courses (Foley-Peres & Poirier, 2008). College math placement test scores proved not to be indicative of specific SAT math scores, even though placement level and math SAT scores seemed to be somewhat positively correlated. In other words, students with higher SAT scores generally placed in higher-level math courses, but numerous exceptions prevailed (Foley-Peres & Poirier, 2008). Faculty members at the college admitted that a few students may have been placed above or below their ability levels, but were given the opportunity to change to a different course (Foley-Peres & Poirier, 2008). Overall, the research revealed that the college math assessment score may be a better indicator of the proper math level than the SAT score (Foley-Peres & Poirier, 2008).

Marwick's (2004) study evaluated four different placement methods: (a) placement by test score alone using ACCUPLACER, (b) placement by high school preparation, (c) placement by considering test score and years of high school math, and (d) placement by student choice after analyzing test score and high school preparation. Although the study found that no one method had a significantly greater impact on academic success, it did show that alternative methods for placement are equally useful (Marwick, 2004; Ruiz, 2007).

A study by Mekonnen and Reznichenko (2007), investigating the reliability of test scores and validity of ACCUPLACER used at Baltimore City Community College (BCCC) as a single tool to place students, concluded that scores of the CPT administered by the college have less consistency (p.11). The CPT can be used to predict academic readiness for a course, but does not guarantee success in the course because of other factors (Mekonnen & Reznichenko, 2007). Final recommendations included the use of students' academic background along with CPT scores to place students in appropriate course level (Mekonnen & Reznichenko, 2007), but stressed the need for further research.

Overall, the above results indicate that alternatives to using one indicator, namely standardized placement score, may be beneficial to students as they result in better placement and better meet students' educational and academic needs (Ruiz, 2007).

College placement test preparation. Many studies claim that most students are unaware and unprepared for the placement test, and consequently enroll in a course that is essentially a review of material previously learned (Ruiz, 2007). Additionally, many students are not aware the enormous impact the placement test has on their college career (Ruiz, 2007). In terms of student awareness, a study by Ruiz (2007) revealed contrary results, namely 72% of students were aware of the placement test, but 85% of students did not study for the test.

To improve student outcomes in the testing process, some institutions have created programs to prepare students for standardized test taking (Ruiz, 2007). Researchers have found statistically significant improvements in standardized test scores as a result of preparation workshops (Ruiz, 2007). These findings are underscored by the results presented by Ruiz (2007), which revealed that students who participated in review courses had a higher mean grade and were more successful in college-level coursework. Another study by BCCC suggested creating preparatory classes for the CPT, i.e. familiarize the student with the test content and urging students to partake in such classes before taking the test, would improve placement outcomes (Mekonnen & Reznichenko, 2007).

Math preparatory workshops provide an excellent opportunity for students to review math content without investing a semester in remedial math and allowing access to college-level math (Ruiz, 2007). Review workshops were particularly helpful to students with a math background and/or for those students who returned to college after a few years since course taking had elapsed (Abraham & Creech, 2000; Ruiz, 2007). As such it helped place students on the path to degree completion without the burden of unnecessary coursework (Ruiz, 2007).

Research by Golfin et al. (2005) revealed that providing applicants with a practice placement exam before the actual placement test reduced the number of students required to take developmental math. Placement scores for those who took the practice exam were significantly higher than for those who did not, and resulted in fewer students being placed into remedial math (Golfin et al., 2005).

In summary, providing students with advanced knowledge of the placement test process and allowing students to prepare for the test could have a positive impact on college-level math placement (Ruiz, 2007).

Retake option and placement appeal. Another approach, namely offering students the option to retake the placement test, received some consideration by community colleges

(Calcagno & Long, 2008). Retake policies vary greatly among colleges. Some colleges allow retake every 90 days without stipulation whereas other colleges allow retake only once a year or no retake (Geraci, 2007). For example, Mesa San Diego Community College allows no retakes under any circumstances; Santa Fe Community College and Naugatuck Valley Community College allow no retakes, although students may appeal this decision based on unusual circumstance; Kapiolani Community College allows a retake every 120 days, but a student cannot take a retake while enrolled in the course; Northern Virginia Community College allows a retake only one time with written permission from a division chair or academic dean, and must be at least one year from the original placement test date; Miami Dade Community College stated that a student can only retake the placement test once prior to enrollment in the placed course; Palm Beach Community College allows students who originally placed into college-level mathematics one retake and no retake for students who placed into remedial math; and at Sinclair Community College students may retake the placement test once every calendar year (Geraci, 2007).

Little research on the efficacy of retake placement tests exists as most studies investigating alternate placement options focus on placement appeal and issuance of waivers for remediation (Hager, 2005; Nickell, 2005) or self-placement option (Dorrity, 2008; Ruiz, 2007). A study conducted by Sawyer and Welch (as cited in Geraci, 2007) investigated the effects of retesting on placement test scores at American college in Iowa. Results indicated modest increases in retake scores, and only students whose initial score was close to college-level placement improved their placement. A study by Geraci (2007) examined the effectiveness of retake placement tests at Broward Community College (BCC) in Florida. At the time the study was conducted, BCC allowed students to retake the mathematics placement test every 90 days

(Geraci, 2007). However, only 2.4% of students who took the placement test once opted to retake the test (Geraci, 2007). Geraci (2007) concluded that allowing students to retest improved retention rates and resulted in a modest increase in scores, particularly for those students whose initial score was close to college-level placement. The research also revealed that the majority of students (66%) who placed in a higher-level course pursuant to retake performed worse in subsequent courses than students who originally placed in the course (Geraci, 2007). The study, however, did not avail information if students reviewed for the retake placement test, the medium used for review, or if students were provided information on topics in need of remediation.

Identifying strength and weaknesses. Most instruments are imperfect for assessing student strength and weaknesses (Prince, 2005). Kirst (1997) stressed that, students who fail the college placement test "should receive specific information on their strengths and weaknesses along with classes to address deficiencies" (p. 4). For example, neither the SAT nor CAL State placement exams tests are designed to diagnose student weaknesses (Kirst, 1997).

Housatonic Community College (HCC) created a new diagnostic placement test that assessed student weaknesses within the sequence of required competencies (Biswas, 2007). Subsequently, the college designed a fast-track self-paced course, which allowed students to progress through the course as they demonstrated competencies at a specific level (Biswas, 2007). Faculty and college leadership claimed that this format would help alleviate some student frustration, save time and resources, and help improve retention and success rates (Biswas, 2007). The results from the first cohort of 43 students were encouraging. About half of the students completed at least 70% of the course, indicating to faculty that students had progressed to higher competencies and lessened students' remedial requirement (Biswas, 2007).

It has been documented that students, in particular those who have mastered the material previously, often get discouraged by having to spend an entire semester in remedial math and drop out of college (Biswas, 2007). Offering intense experience refresher courses, i.e. "boot camps," for such students to brush up on skills and/or overcome deficiencies would seem to better accommodate their needs than repeating an entire semester (Abraham & Creech, 2000; Biswas, 2007; Blair, 2008; Jaschik, 2008; Thompson, 2006).

In lieu of the above result, community colleges should seriously consider implementing placement tests that identify minimal deficiencies and use a new approach to remediation, e.g. identifying and remediating deficiencies, in particular for students who had learned and mastered math in their secondary education (Davis, 1999). Unfortunately, this area has received little attention in the past and more research is needed to substantiate its efficacy (Biswas, 2007).

Informed math self-placement. The most promising assessment and placement technique appears to be a new program designed and implemented by American River College in Sacramento, CA, called Informed Math Self-Placement (Felder et al., 2007). An alternative to traditional placement and designed to determine students' readiness for college-level math, the program allows students to determine their level of math proficiency (Felder et al., 2007). It also initiates a discussion with current and prospective students about college-level math expectations and gets student actively involved in the process and serves as an avenue to identify math standards that link the placement process to math faculty expectations (Felder et al., 2007). Furthermore, the program has numerous other advantages in comparison to other placement programs such as COMPASS (Felder et al., 2007):

1. Self-placement has the potential to allow schools and colleges to communicate standards across districts, serve as a tool to establish statewide standards for entry-

level college work, and provide an impetus for local community college districts and high school districts to cooperate.

2. Self-placement gives student the freedom to test anywhere 24 hours a day, while COMPASS students had to test in the testing center at specified times.
3. The content of the test was developed by the college faculty and matches the classroom content, in contrast to COMPASS. The self-placement test is an un-timed computerized test, which offers each student the same set of questions, i.e. not adaptive like the COMPASS, and hence is a truly standardized computer test. Additionally, the college faculty controls self-placement content and, at any time, can make modifications to the test, a feature not inherent in COMPASS.
4. The self-placement test gives students a preview of the curriculum and helps students decide if they are ready for a particular course. The math department remarked that it is doubtful that the test covered topics broadly enough (only 7-14 questions on the initial test), and adding more questions (20-30 questions) would provide students with a more comprehensive assessment of their skills.
5. The test appears to be much more cost-effective than COMPASS, although the cost-effectiveness of self-placement is not yet known. There is no per-test cost to the institution provided the student takes the test off-campus and 75% of students test from home. The startup costs for self-placement include paying two faculty members to design the test and programmers to develop the online version. Ongoing costs that the college incurs is paying faculty to revise the test and staffing a research office. With COMPASS the college paid each time a student took the test, approximately $50,000 annually, paid administration cost of $100,000 for the assessment center in

addition to annual staff and maintenance cost. In the long run, self-placement appears to be more cost-effective than COMPASS, allows better control of costs, and reduces dependency on a vendor such as COMPASS.

6. With self-placement instructors influence placement by designing the tests, counselors providing guidance, and students making placement decisions. The entire process is inclusive, in contrast to the traditional prescriptive process, involving students in placement decisions.
7. The self-placement program encourages students to review material and retake the test at a later time.
8. Self-placement exposes academic weaknesses and informs students which areas need improvement.
9. Self-placement helped increase math enrollments and students appeared to take math courses earlier in the college course sequence.

Despite its manifold advantages, self-placement has a few drawbacks (Felder et al., 2007):

1. Although not enough data is available at this point to fully evaluate the program, faculty has expressed the need to add more questions to the initial test and the option to view additional questions at the end of the test. And students have requested to see more questions before selecting a course.
2. Results have shown that students are placing themselves into lower level courses under self-placement than counselors did under COMPASS. More than four out of five students placed themselves at the level recommended by self-placement or below.

3. Concern about the validity of the test surfaced since the supervised test conditions of COMPASS are lost and students can take the test unsupervised online from any location.
4. The test does not provide feedback on questions. Students have requested the answers to questions in order to correct their mistakes.

In summary, the self-placement program appears to be the most effective and efficient assessment program designed as of today. Despite its shortcomings, the program provides a solid foundation for developing a more effective assessment process. One feature, which most likely would improve student placement, is a detailed analysis of students' weaknesses with subsequent referral to review modules remediating such deficiencies (Biswas, 2007). The innovative design of the ALEKS CPT and Learning Modules instrument created in 2009 by ALEKS Corporation makes the aforementioned features available. Research is needed to substantiate the efficacy of the new instrument and as such is the topic of the proposed research.

Conclusion. Making placement decisions can be challenging (Morgan & Michaelides, 2005). The high-stakes nature of placement testing underscores the importance of effective and reliable assessment and placement processes (Morgan & Michaelides, 2005). For one, the central focus should be on the development of reasonable performance standards and the identification of corresponding cutoff scores to enable accurate categorization of student performance (Morgan & Michaelides, 2005). Secondly, designing a CPT that covers the broad content area of the course would maximize the validity and reliability of the test and as such the CPT would need to include a sufficient number of questions (Mekonnen & Reznichenko, 2007; Ruiz, 2007). Thirdly, creating preparatory classes for students taking the CPT and encouraging students to take a review class before attempting the test would help familiarize students with the test content and reduce incorrect placement (Golfin et al., 2005; Mekonnen & Reznichenko, 2007; Ruiz, 2007).

Fourthly, colleges should use multiple assessment measures such as students' academic background, ACT, SAT, and AP scores in addition to CPT results to place students in appropriate course levels (ACT 1993; Armstrong, 2000; Marwick, 2004; Mekonnen & Reznichenko, 2007; Ruiz, 2007). Fifthly, creating a test that allows an in-depth identification of deficiencies, in particular for students who previously have mastered higher levels of math, and offering refresher courses, review courses, or self-study courses to remediate such deficiencies would allow students to brush up on skills and achieve a higher level of competency (Abraham & Creech, 2000; Biswas, 2007; Blair, 2008; Davis, 1999; Felder et al., 2007; Golfin et al., 2005; Kirst, 1997; Mekonnen & Reznichenko, 2007; Prince, 2005; Ruiz, 2007; Thompson, 2006). Lastly, explaining to students the consequences of non-credit level coursework and involving students in the assessment process could prove to be most effective in increasing student performance (Felder et al., 2007).

All of the above mentioned strategies offer a great potential for improving the accuracy of placement test results. Even though implementation of a more comprehensive assessment placement procedure would require a greater commitment by colleges in the initial assessment process, the long-term benefits to students, institutions, policy makers, taxpayers, and the nation could prove to be enormous (Brown & Niemi, 2007). It is apparent that the effects of remedial math education reach beyond student levels. Hence, the final section addresses the impact of the remedial math dilemma in a larger context.

Remedial Education in a Larger Context

Our nation has made some advances in preparing students for college and providing access to higher education, but we continue to fall behind other countries in improving college opportunities for our students (Kirst & Bracco, 2004; National Center, 2008). "Whatever lead we enjoyed over other countries has been lost as both our [college] participation and completion rates have declined" (National Center, 2008, p. 25). The 2008 report *Measuring Up* concluded that our higher education performance is not commensurate with the current needs of our society and our economy (National Center, 2008). Even though our nation provides access to higher education for all students, current trends limit our nations' ability to advance the educational attainment of our workforce and citizenry and impede our global competitiveness (Brown & Niemi, 2007; National Center, 2008). The discussion that follows examines social, economic, and political implications of remedial education on college preparation, college access, and college completion.

College preparation. Trends in college preparation including racial/ethnic, income, and state disparities as well as educational gaps will be presented in this section.

Trends. Today high school graduates are more likely to enroll in college compared to earlier decades as over 70% of students go on to postsecondary education (Kirst, 2001; National Center, 2008). But far too many graduates leave high school unprepared to succeed in college-level courses and need remediation (Kirst, 2001; National Center, 2008; Prince, 2005; Ruiz, 2007). Furthermore, more students than in the past fail to graduate from high school or take longer to complete high school (Kirst, 2001; National Center, 2008). "The reduced high school graduation rates decrease the pool of potential college graduates and college-educated workers" (National Center ,2008, p. 5). Reality reflects that education and training beyond high school is a prerequisite for employment that supports our middle-class life (National Center, 2008; Ruiz, 2007).

Racial/ethnic, income, and state disparities. The 2008 report *Measuring Up* concluded that no improvements have been made in increasing the percentage of young adults graduating from high school within four years (National Center, 2008). The nation continues to experience disparities in higher education performance by race/ethnicity, income, and by state of residence (Brown & Niemi, 2007; ICHE, 2008; Kirst & Bracco, 2004; National Center, 2008). For example, high school freshmen in Massachusetts are 17% less likely to enroll in college by age 19 compared to their peers in California, and high school freshmen in Pennsylvania are 12% less likely to enroll than those in South Carolina or Utah (National Center, 2008). A report from the Education Trust showed that socioeconomic status and race/ethnicity do make a difference in terms of the likelihood of a student enrolling in college preparatory courses in high school (Brown & Niemi, 2007; Kirst & Bracco, 2004). Approximately 28% of low-income students enrolled in college preparatory courses compared to 49% of medium-income and 65% of high-income students, and 26% of African American and 23% of Latino students compared to 42% of Asian and 34% of White students pursued such courses (Kirst & Bracco, 2004).

Addressing educational gaps. "Given our relative decline internationally and the gaps in higher education performance within [the United States], no state can afford to maintain the status quo" (Brown & Niemi, 2007; National Center, 2008, p. 7). It is imperative to narrow these gaps as it will improve educational and economic opportunity in those states and for the nation as a whole (Brown & Niemi, 2007; Kirst & Bracco, 2004; National Center, 2008).

College access. Racial/ethnic and socioeconomic developments greatly influence college access and as such will be discussed below.

Racial/ethnic and socioeconomic gaps. Over the past three decades, access to college has been fairly consistent and college attendance has increased for all groups, however gaps in enrollment among racial/ethnic and socioeconomic groups have remained constant (Kirst & Bracco, 2004; National Center, 2008). Students from educationally and economically disadvantaged families enrolled in college at lower rates (Kirst & Bracco, 2004), and this trend is in part due to a lack of college knowledge among students and families of different social classes and racial/ethnic backgrounds (Kirst & Bracco, 2004). For example, 73% of White, 56% of Black, and 58% of Hispanic high school graduates will enroll in college next fall (National Center, 2008). In terms of family income, 91% of high school students from families in the highest income group enroll in college, while enrollment rates for students from middle-income and lowest income families are 78% and 52%, respectively (Kirst & Bracco, 2004; National Center, 2008).

College affordability. This dilemma leads to another trend in higher education, namely the deterioration of college affordability (National Center, 2008). Firstly, college tuition continues to outpace family income (National Center, 2008). "The continuation of trends of the last quarter century [will] place higher education beyond the reach of most Americans" (National Center, 2008, p. 8). Secondly, low- and middle-income families have been experiencing relatively flat or declining family incomes. As a result the financial burden of paying for college costs has increased substantially (National Center, 2008). Thirdly, over the last decade, student borrowing has more than doubled (National Center, 2008). Lastly, students from middle- and upper-income families receive larger grants from colleges and universities than students from lower-income families (National Center, 2008; NCES, 2005).

Although many states have increased the emphasis in need-based financial aid, tuition increases have outpaced growth of financial aid (National Center, 2008). College expenses as a percentage of family income after financial aid is much greater for students from working and poor families, namely 40%, while student from middle- and upper-income families pay 25% and 13%, respectively (National Center, 2008).

Enrollment shifts. Tuition increases are steepest at four-year institutions, but spending on core academic programs and instruction has declined per full-time equivalent student (Delta Project, 2008). A 2008 study *The Growing Imbalance* identified dramatic enrollment shifts of low-income, Black, and Hispanic students from four-year institutions to public two-year and proprietary institutions (Delta Project, 2008). It is unclear whether these enrollment shifts are attributable to rising tuition or to greater competition for limited seats in the four-year sector, or a combination of both (Delta Project, 2008). The dualism of student unpreparedness and the deterioration of college affordability will not only result in putting more pressure on community colleges, but also raises the question of how the U.S. can meet future workforce needs for baccalaureate, master's, and professional degree holders (Delta Project, 2008).

Impact of college unpreparedness. Most students who enroll in community colleges are required to take at least one remedial course (Kirst, 2001, 2003; National Center, 2008; Ruiz, 2007; Smydo, 2008). As such it poses an additional financial burden for those students and families (National Center, 2008; Smydo, 2008). Research has shown that students who begin in remedial courses take longer to complete and have a lower probability of finishing an academic program. Low-income and minority students are affected the most by mandatory developmental course placement (Kirst, 2003; Kirst & Bracco, 2004; Ruiz, 2007). Trends in enrollment shifts are alarming, especially for future generations of first-generation and low-income students, as they most likely will translate to growing social and economic inequality in our society (Delta Project, 2008).

College completion. College completion has never been strength of American higher education (National Center, 2008). In 2008, the U.S. ranked 15th among 29 countries compared in the study by Measuring Up (National Center, 2008). Even though rates of completion have improved slightly over the last four years, they continue to be unsatisfactory (Blair, 2009; National Center, 2008).

Racial and ethnic disparities. Racial and ethnic disparities are not only found in college access but also in college completion and course completion rates (Delta Project, 2008; ICHE, 2008; National Center, 2008). At two- and four-year institutions completion rates range from 42% to 68% (Delta Project, 2008). Of students enrolling in four-year institutions 59% of White, 47% of Hispanic, and 41% of African American students completed a baccalaureate degree within six years (see Table 2) (Delta Project, 2008; ICHE, 2008; National Center, 2008).

In comparison, the overall success rate of students in community college courses appears to be more encouraging, i.e. those students not taking remedial courses. For example, at California community colleges the success rate of students in college-level courses was approximately 64% (Brown & Niemi, 2007). Whites had the highest rate of success at 71%, while those of Hispanic and African American students was below the overall success rate, namely at 61% and 62%, respectively (see Table 2) (Brown & Niemi, 2007; National Center, 2008).

However, rates were much lower for students enrolled in remedial courses and greater variations across ethnicity existed for student success rates in pre-collegiate courses (Brown & Niemi, 2007). For example, at California community colleges 62% of White, 54% of Hispanic, and 41% of African American students completed remedial courses (see Table 2) (Brown & Niemi, 2007). Even though the success rate of students in community colleges is not overwhelming, it appears to mirror the success rates of two- and four-year institutions with the exception of Hispanic students (Brown & Niemi, 2007) (see Table 2).

Table 2

College Completion Rates and Rates of Success by Ethnicity

Ethnicity	College Completion Rates at Two- and Four-Year Institutions	Rates of Success at California Community Colleges	
		College-Level Courses	Pre-Collegiate Courses*
White	59%	71%	62%
Hispanic	47%	61%	54%
African American	41%	62%	41%

* Pre-Collegiate courses are Pre-Collegiate Basic Skills courses commonly referred to as remedial (Brown & Niemi, 2007)

Therefore, it is difficult to argue that students are placed into courses that well match their abilities or achievement levels (Brown & Niemi, 2007). Inconsistent and varied processes for placement testing may contribute to student success and create unnecessary barriers for students entering community colleges (Brown & Niemi, 2007). Specifically, a study by Ruiz (2007) revealed that 70-80% of students, who were considered ineligible for a college-level course, were successful when given the chance to attempt such a course.

Other studies have shown that students are less likely to graduate if they attend on a part-time or intermittent basis, are academically underprepared, work more than 20 hours per week, fall behind in credits in their first year of college study (i.e. students in need of remediation), and/or are first-generation minorities or low-income students (ICHE, 2008; National Center,

2008; Prince, 2005). Those students are traditionally overrepresented in developmental education programs and courses (Prince, 2005; Ruiz, 2007). For example, in California community colleges only one in four degree-seeking students actually earned a certificate or a degree, and the likelihood of taking a transfer level math course after starting in a basic math course was only 10% (Brown & Niemi, 2007). Another example is Indiana community colleges that have a two-year graduation rate of 12% for full-time students seeking an associate degree (ICHE, 2008).

Timely graduation. Timely graduation can benefit the students, the community, the institution, and the state (ICHE, 2008; National Center, 2008; Prince, 2005). Students benefit in terms of savings in total educational expenses; students, communities, and the state benefit from a quicker turnaround into the workforce and the associated earning potential by shortening the time-to-degree; and the institution benefits from a more-efficient use of its capacity of human and physical resources to improve quality (ICHE, 2008). Not completing a two- or four-year college education has substantial negative financial implications to the individual, the state, and the economy (ICHE, 2008; Prince, 2005).

Financial and economic implications. In 2000, a study by the Mackinac Center for Public Policy (2000) reported that remedial education may be taxing the U.S. economy at a rate of 16.6 billion dollars per year (Parr et al, 2006). Most recent findings suggested that the achievement gap costs the country between $3 and $6 billion daily (NCTM, 2009). The consequences imposed on the economy are equivalent to a permanent national recession, and as such it is reflected in lower earnings, poorer health, and higher incarceration rates (NCTM, 2009). According to a 2006 report by the nonprofit Alliance for Excellent Education, community colleges spend $1.4 billion on remedial courses for recent high school graduates

(Smydo, 2008). For example, in Pennsylvania remedial education at community colleges and state system universities cost $34.7 million in 2007-08 (Smydo, 2008).

From a student perspective not completing a two- or four-year college education severely impedes an individual's lifetime earning potential as it is nearly $1 million less than that of an individual with a college degree (ICHE, 2008). A report by ICHE stipulated that ten thousand additional students earning a four-year college degree could add as much as $250 million per year to the economy (ICHE, 2008). Conversely, the benefits of simply attending college without completing are not clear, but it is estimated that an individual with some college may increase their annual earning potential by $4,423 more than a high school graduate (ICHE, 2008).

Policy making. As the cost of education continues to rise and enrollment in remedial education continues to increase, states are wrestling with substantial budget shortfalls and the country's financial crisis (Blair, 2009; Delta Project, 2008; National Center, 2008). States and the nation have no choice but to combat these economic and educational problems (Delta Project, 2008; National Center, 2008). State leaders can respond to the current budget crisis in the usual pattern of the past, namely being passive in allowing the financial burden to be passed on to students and families (National Center, 2008). Consequently our national and state gaps in college access and completion will worsen, and college affordability will continue to deteriorate (Delta Project, 2008; National Center, 2008).

Or responsible state leaders establish state and national policies that remedy the dilemma of our higher education system (National Center, 2008). Much talk exists about creating a seamless transition between secondary and postsecondary education (i.e. K-16 policies providing curricular coherence), demand greater institutional accountability, and improve consumer information (Blair, 2009; Brown & Niemi, 2007; Delta Project, 2008; Kirst, 2001; Venezia,

2000). This approach would address disparities in educational opportunity and achievement, improve preparation for college, reduce the need for postsecondary remediation, reduce time to degree and college completion rates, as well as reduce or at least contain educational expenditures (Abraham & Creech, 2000; Delta Project, 2008; Kirst, 2001). Kirst (2001) noted that, "The governor's office [seems] the logical place for states to align their fractured K-16 standards, but higher education leaders want to guard their political independence from gubernatorial and legislative interference in admission criteria" (p. 17). Hence, it requires a concerted effort by institutional leaders and public policy officials (Brown & Niemi, 2007; Delta Project, 2008; Kirst, 2001; Kirst & Bracco, 2004; National Center, 2008).

Previous attempts at the federal level to bridge the achievement gap between high schools and colleges and bring uniformity to the system have not been successful. For one, policymaking has been relegated to the state level and each state consequently has developed and implemented its own K-12 standards and assessment (Kirst, 2001). "When President Clinton spoke in support of voluntary national testing, he was silenced by protestors championing states' rights, local control of schools, and students' freedom and opportunity to learn" (Kirst, 2001, p. 17). Improving consumer information, i.e. educating parents and students as to how their money is being spent and how much of their money ends up in the classroom, might change institutional behavior (Delta Project, 2008).

Even though "the political and rhetorical boiler plate of K-16 is becoming commonplace," most policy remains within its educational sphere as does funding, accountability, and governance (Delta Project, 2008; Kirst, 2001, p. 6). Perhaps the stimulus for fundamental reform will come from rising public concern about economic, social, and political costs of postsecondary remediation and the demand for more accountability on the part of postsecondary institutions (Delta Project, 2008; Kirst, 2001).

Conclusion. Modest advances in college preparedness, college access, and college completion have been made in recent years, and these improvements are overshadowed by larger gains of other countries and the deterioration of college affordability (Delta Project, 2008; National Center, 2008). "The relative erosion of our national 'educational capital' has occurred at a time when we need more people to be college educated and trained because of Baby Boomer retirements and rising skill requirements for new and existing jobs" (National Center, 2008, p. 9). To increase the educational attainment of our population and the nation's comparative standing internationally, the U.S. must address disparities in educational opportunity and achievement to improve educational and economic opportunity for all Americans and to meet its workforce needs to compete globally (Delta Project, 2008; National Center, 2008).

Summary

This literature review provided a detailed discussion of causal and contextual elements contributing to the remedial math dilemma, an examination of current assessment placement processes, and the ramifications in a social, economic, and political context.

Researchers have taken various perspectives to explain the causes of the remedial math dilemma at the postsecondary level such as: (a) Secondary school curricula do not equip students with the skills needed to perform college-level work (Biswas, 2007; Brown & Niemi, 2007; Kirst, 2003; Kirst & Bracco, 2004; NCES, 2003), (b) the math gap phenomenon (Hager, 2005; Kirst, 2003; Ruiz, 2007; Thompson, 2006), (c) shortcomings in assessment and placement processes and policies and the accountability of test procedures used at community colleges (Armstrong, 2000; Brown & Niemi; ICHE, 2008; Kirst, 1997, 2003; Mekonnen & Reznichenko,

2007; Mitchell, 2006; Prince, 2005; Ruiz, 2007; Venezia, Kirst, & Antonio, 2003), (d) student behavior and attitude (Hager, 2005; Kirst, 2003; Kirst & Bracco, 2004; Ruiz, 2007; Thompson, 2006), and the existing disjuncture between K-12 education and postsecondary education (Brown and Niemi, 2007; Kirst and Bracco, 2004; Kirst &Venezia, 2006).

The majority of studies identified shortcomings in assessment placement processes and the accountability of test procedures used at community colleges as key contributors to this dilemma (Biswas, 2007; Brown & Niemi, 2007; Collins, 2008; Golfin et al., 2005; Kirst, 2001). The increase in enrollment of students considered to be underprepared has led to a growing trend at community colleges to implement a mandatory placement policy and the use of CPTs to determine college readiness of entering students (Collins, 2008; Golfin et al., 2005; Prince, 2005; Shults, 2000). The most commonly used placement tests are standardized norm-referenced tests such as ACCUPLACER, COMPASS and ASSET and a source of controversy (Brown & Niemi, 2007; Gerlaugh et al., 2007; Golfin et al., 2005; Kirst, 2001; Kirst & Bracco, 2004; Ruiz, 2007) for the following reasons: (a) Standardized norm-referenced tests, considered aptitude tests, neglect to measure student achievement based on academic standards (Mitchell, 2006), and (b) the insufficient number of questions given on such tests (Mekonnen & Reznichenko, 2007).

Setting of cutscores poses another challenge and possibly leads to misclassification of students (Mekonnen & Reznichenko, 2007; Ruiz, 2007). Most colleges admit that setting of an optimal cutscore proves to be elusive and a number of institutions use higher cutscores than suggested by the publisher (Mekonnen & Reznichenko, 2007; Ternes, 2009). However, use of higher cutscores most likely results in placing an unquantifiable number of students in courses below proficiency level (Armstrong, 1999; Brown & Niemi, 2007; Golfin et al., 2005) and possibly inflates the number of students placed in remedial math (Ruiz, 2007).

Shults (2000) and Smydo (2008) remarked that assessment and placement of students into remedial courses remains one of the most debated aspects of remedial education. Reports prepared by the U.S. Department of Education and the Achieving the Dream Foundation acknowledged the extent of the assessment placement dilemma and strongly urged colleges to focus more on the initial assessment process to evaluate students and affect more correct placement (Biswas, 2007; Collins, 2008). Various efforts have been undertaken by community colleges to improve student outcomes in the testing process, such as the creation of preparation workshops, review courses or practice exams, to prepare students better for standardized test taking and familiarize students with the test content. Researchers have found statistically significant improvements in standardized test scores as a result of such interventions (Mekonnen & Reznichenko, 2007; Ruiz, 2007).

Another approach, namely offering students the option to retake the placement test, received some consideration by community colleges (Calcagno & Long, 2008). However, little research on the efficacy of retake placement tests exists. Using multiple measures for assessment purposes in addition to using the standardized placement test avails another option to improve placement outcomes. Armstrong (2000), Mekonnen & Reznichenko (2007), and Ruiz (2007) concluded that use of multiple measures may be beneficial to students.

In a larger context the remedial math dilemma has social, economic, and political implications. Even though the United States provides access to higher education for all students, our nation's higher education performance is incommensurate with the current needs of our society and economy for the following reasons (National Center, 2008): The United States (a) continues to experience disparities in higher education access and performance by race, ethnicity, income, and by state of residence; (b) exhibits persistent gaps in enrollment among racial/ethnic and socioeconomic groups; (c) experiences a deterioration of college affordability; (d) continues to fall behind other countries in improving college opportunities for our students; and (e) has lost the lead over other countries in college participation and completion (Brown & Niemi, 2007; Kirst & Bracco, 2004; National Center, 2008; Prince, 2005). These trends limit our nation's ability to advance the educational attainment of our workforce and citizenry to meet future workforce needs (Delta Project, 2008) and impede our global competitiveness (Brown & Niemi, 2007; National Center, 2008).

Brown and Niemi (2007) stipulated that inconsistent and varied processes in placement testing create unnecessary barriers for students entering community college, negatively impact student success and completion rates, and have substantial negative financial implications to the individual, the state, and the economy. Astin (2002) advised that, "assessment policies and practices in higher education should always give full consideration to the evaluative uses to which our measures will be put" (p. 3). Due to the high stakes attached to placement decisions, it is imperative that the placement process be as solid and defensible as possible (Morgan & Michaelides, 2005). Research findings point to an inherent dichotomy in the assessment process at most community colleges and subsequent placement of students. The intent of assessment is to fairly place students in appropriate courses and to help students gain access to college-level math courses as quickly as possible. But many of the tools and strategies used in the assessment placement process do not seem to produce the intended results (MeKonnen & Reznichenko, 2007).

Improvements in assessment placement processes and policies would not only benefit students and parents, but also institutions, the taxpayer, and the nation (Brown & Niemi, 2007). Prince (2005) remarked that, "these benefits would accrue to all students who take developmental education courses, but they would be particularly important to low-income, minority, and first-generation college students who are traditionally overrepresented in developmental education programs and courses" (p. 2).

In view of current trends in remedial math education at community colleges, i.e. the continually increasing number of students placed in remedial math and the inherent shortcomings in the assessment placement process, and the country's financial crisis which has led to substantial budget shortfalls for higher education institutions, improvements of assessment placement processes should receive highest priority (Collins, 2008; Kirst, 2001). Further research investigating assessment processes and intervention strategies to improve student outcomes is warranted. Hence, the proposed study added to the body of knowledge and possibly helped students reach higher levels of success.

Chapter 3: Research Method

A new approach to math assessment placement for entering community college students deemed warranted. Numerous researchers suggested that colleges focus on identifying and remediating specific deficiencies for students who had learned higher-level math in their secondary education, e.g. targeted remediation, and offering students the opportunity to retake the placement exam (Abraham & Creech, 2000; Biswas, 2007; Calcagno & Long, 2008; Davis, 1999; MeKonnen & Reznichenko, 2007; Ruiz, 2007; Thompson, 2006). Currently, students who fail the computerized placement test (CPT) are not provided specific information on what to review to address their deficiencies (Kirst, 1997) and offered the opportunity to retake the CPT (Calcagno & Long, 2008; Geraci, 2007).

Most placement testing instruments are imperfect for assessing student strengths and weaknesses (Kirst, 1997; Prince, 2005). Results of one-on-one interviews with students, who successfully completed college preparatory math courses in high school, revealed that many students were disappointed with their placement. Students claimed they could achieve higher placement if they were given a list of topics to review (Ruiz, 2007). Few research studies have been completed on the efficacy of retake policies (Calcagno & Long, 2008; Geraci, 2007). Virtually no attention has been devoted to identifying specific deficiencies and integrating targeted remediation into the assessment and placement process to improve placement outcomes and reduce remediation rates (Abraham & Creech, 2000; Biswas, 2007; Prince, 2005; Thompson, 2006).

The purpose of this quasi-experimental quantitative study was to explore the efficacy of implementing targeted math remediation into the assessment placement process to improve test scores, placement outcomes, and success rates of entering community college students. A quantitative methodology best met the needs and purpose of the research study as it allowed investigation of the relationship between targeted math remediation, the independent variable, and student outcomes, the dependent variable (Larson & Farber, 2003). This approach helped provide a better understanding of how to reduce remediation rates as a result of more accurate placement, and in turn, possibly improve college success and degree completion rates. The proposed intervention offered students the opportunity to remediate specific math deficiencies identified via the initial CPT and retake the CPT to improve placement outcomes. The following research questions and associated null (H_0) and alternative (H_a) hypotheses guided the proposed study.

Q1. To what extent did targeted math remediation during the assessment placement process affect CPT scores?

$H1_0$: There was no statistically significant difference in student initial and retake CPT scores for students who engaged in targeted math remediation.

$H1_a$: There was a statistically significant difference in student initial and retake CPT scores for students who engaged in targeted math remediation.

Q2. To what extent did targeted math remediation during the assessment placement process affect student placement?

$H2_0$: There was no statistically significant difference in student placement outcomes for students who engaged in targeted math remediation.

$H2_a$: There was a statistically significant difference in student placement outcomes for students who engaged in targeted math remediation.

Q3. To what extent did targeted math remediation during the assessment placement process affect course completion rates?

$H3_0$: There was no statistically significant difference in student success rates for students who placed in college-level math as a result of the intervention compared to students who originally placed in the course.

$H3_a$: There was a statistically significant difference in student success rates for students who placed in college-level math as a result of the intervention compared to students who originally placed in the course.

These questions were derived from the possible misplacement of students into remedial math at community colleges, in particular those students who had previously learned higher levels of math (Golfin et al., 2005; MeKonnen & Reznichenko, 2007; Ruiz, 2007). The objective of this research was to investigate a possible intervention strategy to reduce the number of students placed in remedial math. The final aspect of this research was to understand whether targeted remediation during the assessment placement process would help students reach higher levels of success.

The chapter provides a description of the research methodology and design implemented, participants involved in the study, the setting selected for the study, and the instruments used to assess student proficiencies. Operational definition of variables illuminates how data gathered related to the construct of the study. The final section avails an in-depth account of data collection, processing methods, and analysis tools used to prove or disprove the claim stipulated in the research questions as well as a discussion of study limitations and ethical considerations.

Research Methods and Design

A quasi-experimental quantitative study best met the needs and purpose of the research in studying the efficacy of the proposed intervention, a cause-effect relationship (Trochim & Donnelly, 2007). This approach allowed investigation of the relationship between targeted review, the independent variable, and student outcomes, the dependent variable. The basic design followed a repeated-measures design implying that the same measure, the CPT, was administered before and after the intervention (Coladarci et al., 2008). Examination of alternate research designs substantiated this claim and discussion of such is provided below.

Alternative research designs. A qualitative approach to the proposed study using semi-structured, one-on-one interviews most likely would have aided in identifying a phenomenon and developing a theme. In this case, gain a thorough understanding of participants' perceptions and attitudes regarding the feasibility of incorporating a new intervention strategy into the assessment placement process (Eysenck, 2004). A mixed methods research methodology, such as concurrent transformative mixed methods design incorporating semi-structured interviews and quasi-experimental design appealed from a theoretical point of view (Tashakkori & Creswell, 2007).

However, qualitative and mixed methods designs inherit significant theoretical and practical limitations. The intent of the proposed research was to produce quantitative evidence with respect to the efficacy of the intervention. In this case, quantitative variables such as test scores, course placement, and success rates were used to operationalize the research questions. Qualitative research tends to be more general and exploratory than deterministic (Eysenck, 2004). Studying students' and teachers' perceptions and attitudes would not have aided in fulfilling the premise of the study. Hence, research that incorporated one-on-one or semi-structured interviews, such as qualitative and mixed methods designs, would not have been conducive to studying the efficacy of the intervention in terms of quantitative measures.

Furthermore, conducting interviews with students and instructors stipulated that the researcher be physically present at the testing center as students complete the testing process and teachers are available to engage in interviews. This would have required multiple visits to the campus and limited the practicality of conducting research given the researcher's need to engage in long distance travel to conduct interviews, i.e. requiring time and financial resources not available. From a practical and theoretical point of view, qualitative and mixed methods research methodologies did not meet the needs and purpose of the research.

Optimal research design. A one-group pre-post quasi-experimental design availed the optimum choice for the proposed research and best met the needs and purpose of the proposed study considering the intent and circumstances. Firstly, the design was conducive to investigating the cause-effect relationship stipulated and examining the effectiveness of the intervention from a statistical perspective. Secondly, access to the campus database availed a large data set warranting a high degree of confidence in measurement of treatment effectiveness. Thirdly, even though a quasi-experimental design inherited methodological weaknesses, validity of data was not likely undermined provided weaknesses were recognized and accounted for during the experimental process. Shuttleworth (2008) noted that this type of study "can be a very powerful tool, especially where a true experiment is not possible" to help improve students' lives and contribute to society (p. 1). In this case, providing students the opportunity to remediate deficiencies prior to final placement to reach higher levels of success promoted empowerment and liberation. In summary, the study's goals and purposes were best met via a quantitative study design.

Participants

Participation in the study was limited to students who did not meet the requirements to enroll in college-level math, i.e. scored below the cutscore, and volunteered to review material prior to retaking the placement test. The cutscore was used to identify students ineligible to enroll in college-level math and thus eligible to participate in the program. Students self-selected to participate in the program through guidance of campus personnel, e.g. testing personnel and counselors.

Power analysis availed a tool to determine the sample size needed to assure accuracy of conclusions about treatment effects and was an intricate part of experimental design (Houser, 2007). Prospective power analysis via G*Power analysis software indicated a required sample size of 90 for a two-tailed *t* test with two dependent samples. Calculations were based on the following assumptions: (a) an effect size of $\rho = .30$ or larger due to a low-to-moderate correlation as a result of possible restriction of range inherent in data collected, (b) a desired level of power at .80 per Cohen's (as cited in Coladarci et al., 2008) proposed convention, and (c) a significance level $\alpha = .05$. G*Power analysis for ANOVA: Fixed effects, omnibus, one-way, estimated the following priori sample size to achieve a desired power of .80, alpha = .05, and a medium effect size ($\rho = 0.30$ or less) based on sensitivity analysis for gender, age, and ethnicity: (a) a sample size of 150 for groups = 2 and an effect size = .2303, (b) a sample size of 150 for groups = 5 and an effect size = .2867, and (c) a sample size of 186 for groups = 6 and effect size = .2667.

Setting the effect size to $\rho = .30$ or larger specified the smallest population effect size the researcher wanted to be able to detect (Coladarci et al., 2008). With a power of .80, the probability is .80—eight in ten—that the statistical test uncovered a population effect size of $\rho =$ .30 or larger (Coladarci et al., 2008). In other words, in eight of ten cases there was an effect where the alpha level also determined the likelihood of a false positive (Trochim & Donnelly, 2007). A sample size of 172 sufficed to validate results. Conclusions about treatment effects were most likely accurate and indicative of true effects.

Southwest Community College (SWCC) was selected for this study due to its pioneering efforts in assessment placement practices, namely offering students the opportunity to remediate specific deficiencies identified in the initial CPT and to retake the CPT. The research site was a single college in a large multi-college district in the Southwest. The institution and district were given fictional names Southwest Community College (SWCC) and Southwest District (SWD), respectively. All 10 colleges in the district, with the exception of SWCC and two other colleges, used ASSET, COMPASS, and ACCUPLACER to assess the skill level of students in mathematics.

In 2010, SWCC decided to pilot the newly developed placement testing instrument ALEKS (ALEKS, 2010). With this SWCC constituted one of few colleges in the United States, if not the only college, to have utilized the testing instrument at the time. Due to the college's progressive efforts in improving assessment placement processes, SWCC served as a testing ground for investigating the efficacy of incorporating targeted review into the assessment placement process. The Fall 2009 student demographics for SWCC and SWD are shown in Table 3 (MCCCD, 2010).

Table 3

Student Demographics for SWCC and SWD

Fall 2009	SWCC	SWD
American Indian	3%	3%
Asian/Pacific Islanders	5%	4%
African American	4%	7%
Hispanic	17%	20%
White	63%	55%
Other	8%	11%

Instrument

A new testing instrument ALEKS, a commercially developed CPT by ALEKS Corporation, made testing of the intervention possible (ALEKS, 2010). ALEKS is an acronym for Assessment and LEarning in Knowledge Spaces (ALEKS, 2010). ALEKS, the practical realization of Knowledge Space Theory, is the result of ground-breaking research in mathematical cognitive science initiated by Professor Jean-Claude Falmagne at New York University and the University of California, Irvine, and Professor Jean-Paul Doignon at the University of Brussels (ALEKS, 2010). The ALEKS Math Placement Test is a web-based, artificially intelligent assessment and learning system that zeros in on the strengths and weaknesses of a student's mathematical knowledge (MCCCD, 2010). The instrument not only assesses student proficiencies, but also performs an in-depth analysis of student content knowledge and provides an individualized detailed report listing topics in need of remediation (Falmagne et al., 2003). Furthermore, ALEKS Learning Modules, a remediation tool, is available for students to engage in target remediation (ALEKS, 2010).

ALEKS Math Placement Test incorporates two separate tests to accurately assess student current math knowledge and to provide appropriate review topics. ALEKS Level I Math Test can place students into Basic Arithmetic, Introductory Algebra, and Intermediate Algebra. ALEKS Level II Math Test can place students into Intermediate Algebra, College Mathematics, College Algebra/Functions and above. For the purpose of this study only Level II Math test results were utilized as such identified students most likely to benefit from the intervention.

The ALEKS system, a standardized criterion-referenced test (CRT), is fully automated incorporating open response (not multiple choice) and adaptive assessment. Students were not able to check previous answers. The length of the assessment was approximately 20-35 questions. The exact number of questions varied due to the adaptive mechanism, i.e. the assessment focused on questioning in an individually appropriate way. Students who did not have enough time to finish the test could log out and come back to the Testing Center at another time to finish. No score was recorded until the student completed the test. Students had six weeks to complete the assessment from the time they began the test. At the end of the test students were provided the following information: (a) the course for which a student has shown readiness to register, and (b) a list of topics the student has mastered and a list of topics to review. Free sample tests for Beginning Algebra, Intermediate Algebra, College Algebra, and Pre-Algebra were available, but students were not able to see which problems they answered incorrectly. In order to review for the ALEKS Math Placement Test, ALEKS tutorials were available online for purchase (ALEKS, 2010).

Students who did not obtain the required numerical scores for the math course they desired to take had the option to retake the SWD ALEKS Math Placement Test in an attempt to improve the score. The college strongly suggested review of topics identified as deficient on the initial test and as such helped guide students in the review process. The college paid for two math placement tests. Students could only take two ALEKS Placement Tests per year (MCCCD, 2010).

Operational Definition of Variables

Operational definition of variables was instrumental in establishing the connection between construct and operationalization. The intent of this study was to gather data on students' educational performance and demographics to support the hypothesized claims, i.e. targeted remediation increased test scores and improved student outcomes. The study examined if students who scored below the cutscore on the initial ALEKS Math Placement Test and offered the opportunity to review material and retake the placement test would improve test scores, placement outcomes, and success rates. One specific indicator, namely targeted remediation via ALEKS Learning Modules (ALEKS, 2010), was identified to operationalize the research questions (Siniscalco & Auriat, 2005; Trochim & Donnelly, 2007). Even though targeted math remediation could not be measured explicitly, placement test scores, course placement, and course completion rates allowed investigation of the efficacy of the intervention strategy. To operationalize the construct, data pertaining to the following variables was extracted from the SWCC database: (a) Initial CPT scores (IT) and retake CPT scores (RT); (b) course placement subsequent to retake (RP); (c) course grade (GR); and (d) demographics such as age (A), gender (G), and ethnicity (E). Reasoning for inclusion and construct of each variable are provided below.

CPT scores. Initial (IT) and retake (RT) placement test scores, variables numeric in nature, measured student knowledge. Scores extracted from the college database ranged from 0 to 100. Data gathered via the database provided evidence as to whether students successfully improved their math proficiency given the opportunity to review and retake the CPT, and were instrumental in proving or disproving the claim stipulated, namely students who engaged in targeted review during the assessment placement process increased placement test scores.

Course placement. Retake placement (RP) represented student placement pursuant to the intervention. Placement was defined as remedial math (e.g. Arithmetic Review, Beginning Algebra, Math Concepts and Applications, and Intermediate Algebra) and college-level math (e.g. College Mathematics, College Algebra, and Trigonometry/Pre-Calculus) coded 0 and 1, respectively. The proportion of students placed in college-level math subsequent to the intervention, a numeric measure, was calculated and segmented into smaller groups based on demographics. Data gathered via the college database provided evidence to prove or disprove the claim stipulated, namely students improved placement, e.g. achieved a score above the cutscore given the opportunity to review and retake the placement test.

Success rate. Success rate (SR), a variable numeric in nature, was determined for students who gained access to college-level math as a result of targeted remediation and retaking the placement test and enrolled in the course. Success rates were defined as the proportion of students who successfully completed a college-level math course compared to the total number of students enrolled, and did not include students who withdrew or received an incomplete. Grades were coded as follows: Grades A, B, and C = 1 and grades D and F = 0. Data gathered via the college data base provided evidence to prove or disprove the claim stipulated, namely success rates for students enrolling in college-level math pursuant to remediation were equal to success rates of students originally placed in the course. Data was also segmented into smaller groups based on demographics to determine differences between groups.

Related literature. Some variables used in this study have been incorporated in other research. Most of the studies examining math placement issues were quantitative in nature and explored constructs such as cutscores, CPT scores, success rates, student performance, impact of review courses and practice exams on test scores, and use of multiple measures.

Several studies on cutscores examined the validity of placement test scores with respect to success rates (Armstrong, 2000; Ruiz, 2007). Other researchers investigated the impact of relaxed cutscores (Ruiz, 2007; Nickell, 2005) and the effect of placing students on the edge of remediation in remedial courses on student performance and success rates (Ternes, 2009). Geraci (2007) and Dorrity (2008) studied the efficacy of retake options on student success in developmental math courses. Numerous other studies examined success rates of students placed in remedial math and the use of multiple measures for student placement (Armstrong, 2000; Geraci, 2007; Hager, 2005; Ternes, 2009). Few of the aforementioned studies used demographic variables to determine differences between groups.

In-depth review of literature produced few results on efforts made by community colleges to encourage students to review material prior to taking the CPT. Often authors alluded to possible reasons why students were unsuccessful in gaining access to college-level math courses, e.g. students did not take the test seriously, students did not prepare for the test, students were not aware of the importance of the test, etc. (Armstrong, 2000; Hager, 2005; Ruiz, 2007). One qualitative study by Hager (2005) suggested that students would be better served receiving counseling prior to taking the placement test and encouraging review of material. The researcher based her recommendation on results of prior studies, which revealed that a small amount of studying prior to taking the placement test resulted in significant higher placement (Ruiz, 2007). In her final remarks, Hager voiced that community colleges provide a disservice to students requiring "students to take an entire course if all they truly needed was a refresher of the material" (p. 25), and stressed the need for additional research investigating the abovementioned intervention.

Data Collection, Processing, and Analysis

A description of data collection and processing methods as well as statistical analysis tools used to determine the efficacy of the intervention are provided.

Data collection. Pursuant to SWCC IRB approval, data for students who had taken the ALEKS Level II Math Test during the first year of implementation was compiled. A total of 1096 students took the ALEKS Level II math placement test with 821 students (75%) scoring below the cutscore set for admission to college-level math. Of the latter group, 172 students

(21%) opted to engage in targeted math review and retake of CPT. This group of students constituted the sample size for this study.

Processing. Students initial and retake placement test scores (ranging from 0 to 100), course placement (remedial math = 0 and college-level math = 1), and course grade (grades A, B, and C = 1 and grades D and F = 0) were recorded in a spreadsheet. The importance of this portion of the research was based on the placement test scores, course placement, and the success rate of students enrolled in the course. Data was purged of personal identifiers and replaced with an arbitrary code.

Analysis. Descriptive measures such as frequency distributions, proportions, success ratios, summary statistics, graphs and charts were used to organize, summarize, and analyze data. "Descriptive statistics describe the basic features of the data in a study and help summarize large amounts of data in a sensible way" (Trochim & Donnelly, 2007, p. 264). However, a comprehensive analysis of data requires use of multiple descriptors or statistics and, in many instances, the use of inferential statistics (Trochim & Donnelly, 2007). Univariate, bivariate, and inferential statistics such as one-sample *t* test, one-sample *z* test for proportions, two-sample t-test for dependent samples, coefficient of variation, ANCOVA, and logistic regression were utilized to address the research questions. The hypothesis was accepted or rejected using accepted statistical practices. Consistent with good practice, all statistical tests were followed up by constructing a 95% confidence interval (CI) (Coladarci et al., 2008). In the event that hypothesis testing resulted in rejection of H_0, a 95% CI should not span the hypothesized value of H_o (Coladarci et al., 2008). All data was analyzed using Excel, StatPlus, and SAS statistical software.

Research question 1. One could reasonably expect that students who engaged in review increased their score on the retake CPT. The following alternative hypotheses were developed for research question one:

H1A: There was a difference in initial and retake placement test scores for students who engaged in independent review or $\mu_{IT} - \mu_{RT} \neq 0$.

H1B: The mean retake score was significantly greater than 37 or $\mu_{RT} > 37$.

H1C: There was a significant difference among gender groups in terms of the mean difference in initial and retake test scores.

H1D: There was a significant difference among ethnic groups in terms of the mean difference in initial and retake test scores.

H1E: There was a significant difference among age groups in terms of the mean difference in initial and retake test scores.

Two-tailed t test for two dependent samples. The "before-after" scenario stipulated use of two variables, the initial CPT score (IT) and the retake CPT score (RT) to test alternative hypothesis H1A (Coladarci et al., 2008). Pursuant to verifying the assumptions of normality of test scores and the requirement of homogeneity of variances (Coladarci et al., p. 167), a two-tailed *t* test for two dependent samples was chosen to test the non-directional alternative hypothesis H_{1A}: $\mu_{IT} - \mu_{RT} \neq 0$, i.e. the difference in means of IT and RT, at a level of significance $\alpha = .05$ (Coladarci et al., 2008; Larson & Farber, 2003). The null hypothesis stated no difference, H_o: $\mu_{IT} - \mu_{RT} = 0$. As the expected number of participants materialized and resulted in a large sample ($n > 30$), the concern inherent in a paired sample *t* test due to a small sample size, namely the impact on the decision to reject and retain a false H_o if Pearson r is low, was nullified (Coladarci et al., 2008).

The *t* test for dependent samples was followed up by constructing a 95% CI for $\mu_{IT} - \mu_{RT}$ (Coladarci et al., 2008). Given the large n, the width of the CI was expected to be fairly narrow. In the event that hypothesis testing resulted in rejection of H_o, the 95% CI would not span zero. The researcher could be 95% confident that the true difference was anywhere from a slight to a somewhat large increase in scores (Coladarci et al., 2008).

Coefficient of variation. The coefficient of variation, R^2 and a measure of effect size, was computed for variables IT and RT and as such aided in better communicating the magnitude of association between IT and RT. This measure allowed estimation of the magnitude of other factors that possibly affected the association of variables (Coladarci et al., 2008). In terms of the proposed study, R^2 allowed determination of the proportion of variance or change in test scores due to review and the proportion due to other factors, i.e. motivation, study habits, hours studied, etc.

However, a word of caution was warranted employing this measure. For one, large samples are more likely to give statistical significant results (Coladarci et al., 2008). Due to the effect that the sample size has on reducing the standard error, a small or even trivial difference in means can result in rejection of H_o when samples are large. However, rejection of H_o does not necessarily imply that the difference is large or important (Coladarci et al., 2008). Secondly, Coladarci et al. (2008) remarked that restriction of range may result in a lower correlation coefficient and an even lower R^2. Study participants constituted only those students who volunteered to engage in independent review and retake of CPT, thus possibly resulting in restriction of range, a lower correlation coefficient and reduction in reliability of results. In this case, Pearson r would reveal little about the relationship between initial and retake test scores.

Subsequent analysis of grouped data, i.e. grouping students by age or sex, may show greater variation (Coladarci et al., 2008).

Close examination of results, i.e. use of effect size to determine the magnitude of difference, and careful inspection of restriction of range and linearity as well as variances, standard deviation, and scatterplots for outliers was warranted prior to making a final conclusion. In addition, description of factors or variables incorporated in the study, tools utilized to measure such variables, and the context in which those measures had been obtained would help explain a low association (Coladarci et al., 2008).

One-sample t test. A score of 38 or higher indicated placement in college-level math (see Appendix B) (MCCCD, 2010). As the population variance was unknown, a one-sample *t* test (α = .05) was used to test the directional alternative hypothesis H1B: $\mu_{RT} > 37$.

ANCOVA. ANCOVA availed a statistical tool to analyze whether mean test score differences for differences in IT and RT scores existed between groups, i.e. gender, ethnicity and sex, provided the underlying assumptions for the F-test were met. Assumptions for a F-test are (a) samples are randomly selected from a normal population, (b) samples are independent, and (c) each population has the same variance (Coladarci et al., 2008; Larson & Farber, 2003). In particular, an ANCOVA Proc GLM model with covariate IT was used to obtain a factorial analysis of variances to compare means of the dependent variable, differences in CPT scores, for groups defined by demographic variables to test the non-directional alternative hypotheses H1C, H1D, and H1E at α= .05 (Norusis, 2006). Significant F ratios warranted examination of least squares means estimates to determine subgroups of demographics more likely to improve test scores (Coladarci et al., 2008).

Research question 2. As students were expected to benefit from the intervention, i.e. improve test scores, research question 2 investigated the effect of such on placement outcomes, i.e. proportions of students placed in college-level math. Improving test scores may or may not result in higher placement, i.e. gain access to college-level math courses, despite an increase in test score. Placement was determined by cutscores set by the college and the publisher (ALEKS, 2010). A retake score above 37 indicated placement in college-level math (see Appendix B). The following alternative hypotheses were developed for research question 2:

H2A: The majority of students who engaged in independent review placed in college math or $p_{RP} > .5$.

H2B: There was a significant difference among gender groups in terms of placement outcomes.

H2C: There was a significant difference among ethnic groups in terms of placement outcomes.

H2D: There was a significant difference among age groups in terms of placement outcomes.

One-sample t test. A score of 38 or higher indicated placement in college-level math. A one-sample *z* test for proportions ($\alpha = .05$) was used to test the one-tailed (directional) alternative hypothesis H2A: $p_{RP} > .5$.

Logistic regression. Logistic regression was utilized to test the effect of demographics on placement outcomes, i.e. test hypotheses H2B, H2C, and H2D. Binary logistic regression was used to predict the binary dependent variable RP coded 0 or 1 on the basis of continuous and categorical independent variables with age a continuous variable and gender and ethnicity categorical variables.

Binary logistic regression is a form of regression, which is used when the dependent variable is a dichotomy and the independent variables are of any type (Garson, 2010). Logistic regression applies maximum likelihood estimation to estimate the odds of a certain event occurring. Logistic regression does not assume linearity of relationships between the independent and dependent variables and does not assume homoscedasticity (Garson, 2010). It does, however, require that observations be independent and that the independent variables are linearly related to the logit of the dependent (Garson, 2010). To test the overall fit of a binary logistic regression model, Garson (2010) recommended Hosmer and Lemeshow Goodness-of-Fit Test (H-L). Non-significance corresponds to an adequate fit of data, i.e. a goodness-of-fit statistic greater than .05 implies that the model fits the data at an acceptable level (Garson, 2010). However, "this does not mean that the model necessarily explains much of the variance in the dependent, only that however much or little it does explain is significant" (Garson, 2010, p. 10).

Research question 3. Research question 3 examined success rates of students who placed into college-level math as a result of the intervention compared to the overall success rate (SR) of .70 for students at SWCC who originally placed in the course. The following alternative hypotheses were developed for research question three:

H3A: Students who placed in college-level math based on CPT retake were not as successful than students who originally placed in the course or $p_{SR} \neq .70$.

H3B: There was a significant difference among gender groups in terms of success rates.

H3C: There was a significant difference among ethnic groups in terms of success rates.

H3D: There was a significant difference among age groups in terms of success rates.

One-sample z test for proportions. The overall success rate of students at SWCC enrolled in college-level math was 0.70. A non-directional alternative hypothesis was chosen to test H3A: $p_{SR} \neq .70$ at a $\alpha = .05$ significance level.

Logistic regression. Logistic regression was employed to test the effect of demographics on success rates, i.e. test hypotheses H3B, H3C, and H3D. In this case, the binary logistic model predicted the odds of the binary dependent variable SR on the basis of the continuous independent variable age and the categorical independent variables gender and ethnicity.

Methodological Assumptions, Limitations, and Delimitations

For any study possible threats to validity exist (Trochim & Donnelly, 2007). These threats present reasons for the conclusions or inferences reached in the study to be considered invalid and can undermine conclusion validity (Trochim & Donnelly, 2007). To increase the quality of research conclusions and to reduce the plausibility of most likely threats to validity, issues and/or variables that possibly influenced or alternative explanations that could possibly be formulated were identified and discussed (Trochim & Donnelly, 2007). Identification of study limitations helped render the credibility of validity of conclusions as well as discredit alternative explanations posed by critics (Trochim & Donnelly, 2007). In terms of the proposed study a number of possible limitations were identified. Strategies for minimizing the impact of such are delineated.

Internal Validity. The proposed study employed a quantitative approach, thus issues relative to validity and reliability needed to be addressed.

Instrument bias. The ALEKS CPT, a new commercially developed placement test, was used to assess student math proficiencies and determine test scores. The test was a validated instrument and had been on the market since 2009 (ALEKS, 2010). SWCC was the first college to use this test for assessment placement and remediation purposes. Testing personnel at the college had voiced concern about the appropriateness of cutoff scores set by the publisher and anticipated that adjustments most likely would be necessary in the future.

Sampling. Since all data collected during the first year of implementation, or a large random sample thereof, was used in the study, the college anticipated that research findings would aid in determining whether the new instrument accommodated the needs of their students and/or college.

Data collection. Any study risks human error. Such errors may range from how the data is gathered to clerical errors in data entry. Steps were taken to minimize data entry errors by screening data for egregious errors. Inquiry as to how the data was collected and the criteria for recording data were also made.

Population sample. Participant selection may have impacted internal validity for the following reasons. For one, participants volunteered for the program solely on the basis of the cutscore on a pre-program measure, i.e. students who scored below the cutoff score on the initial CPT. Research has shown that, "there are systematic personality differences between volunteers and non-volunteers" (Eysenck, 2004, p. 5) and as such may impact reliability and validity of results. Secondly, participants were selected because they possessed some extreme characteristic, namely unpreparedness for college-level math. Hence, differences between pretest and posttest scores could be attributed to regression toward the mean, namely the tendency of extremes to revert toward averages (Coladarci et al., 2008). Thirdly, a large sample, i.e. 300 or more, is required to substantiate significance of results for logistic regression (Hsieh, Bloch, & Larsen, 1998). A sample size significantly below this threshold may impact the reliability and validity of results.

Confounding. Confounding was a concern in this study. Extraneous variables may have influenced the outcome and produced random and/or constant error effects (Eysenck, 2004).

Random error. Individual students' circumstances may have affected the study's outcomes. For example, when taking or retaking the test, the student may have had little sleep, personal problems, be ill, or involved in a major life crisis. The large sample size of the study helped control the effect of individual variables of this nature as the number of students who faced such circumstances should have been relatively small compared to the overall sample size.

Constant error. Constant error is present when the effects of any unwanted variable on the dependent variable differ between conditions or participants (Eysenck, 2004). For example, student motivation, amount of time studied, medium used for review, etc. very likely influenced testing outcomes and a student's decision to engage in review. However, such variables were not addressed in the study design. Further research that includes such variables is recommended.

Content validity and history. Some variables used in this study have been incorporated in other studies, such as CPT scores, placement outcomes, and success rates. An extensive literature review revealed how other studies have explored such variables. However, the independent variable, i.e. targeted review, had not been used in any other study since ALEKS CPT was the first testing instrument that availed an in-depth analysis of students' math competencies and provided a detailed list of topics for students to review to improve placement outcomes. This list was utilized for targeted review. Other commercially developed and widely used CPTs, such as ACCUPLACER, COMPASS, and ASSET, merely offered a test score and were not suitable for targeted remediation. In terms of the proposed study, statistical analysis results provided insight into whether the constructed operationalization, i.e. targeted math remediation, produced the intended results, i.e. improve student outcomes (Trochim & Donnelly, 2007). Further research is highly recommended to test this type of intervention in other settings.

External validity. The research site chosen for the proposed study, a large community college in the Southwest, was a federally designated minority serving institution with a nontraditional student population (Ruiz, 2007). The demographics of the student population made it feasible to generalize results to other community colleges across the nation as low-income and minority students are generally overrepresented at community colleges (Prince, 2005; Ruiz, 2007).

Ethical Assurances

All data collected relating to this study was approved by Northcentral University Institutional Review Board (IRB) and SWCC IRB prior to data collection. No physical or verbal contact was established with human subjects. Informed consent was not an issue in this research as all data was obtained directly from the SWCC database and publicly accessible websites. In this particular study the need to identify individual participants did not exist. To maintain participants right to privacy, each student record was assigned a code number and no names were associated with any of the data collected or any aspect of the study. The reported findings were written to avoid any connection to individuals or groups of individuals that participated. Data transferred from the college database will be kept in a secure place, i.e. locked in a filing cabinet at the researchers home and transferred from the computer hard drive to an external hard drive. Hence, in the event of computer piracy the data would not be accessible. Thus, anonymity of participants was ensured and students were protected from harm. As there was only one researcher for this study, the data and information collected was not shared with other researchers or colleagues. Honesty in communicating results, methods, and procedures with professional colleagues was of utmost importance. Ethics were the most important part of the study and the highest ethical standards were upheld in every aspect of the research and reports.

Summary

To examine if a new approach to math assessment placement at community colleges improved student outcomes, i.e. incorporation of targeted review and CPT retake option into the assessment placement process, was the purpose of this study. The proposed study sought to investigate the relationship between targeted review, the independent variable, and student outcomes, the dependent variable and followed a repeated measures design. A one-group pre-post quasi-experimental design was used to examine the efficacy of the new intervention strategy and best met the study's goals and purposes. The testing instrument used in the proposed research, i.e. the standardized criterion-referenced ALEKS Math Placement Test, was the first testing instrument that availed an in-depth analysis of students' math competencies and provided a detailed list of topics for student to review.

SWCC was selected for the study due to its pioneering effort in offering students the opportunity to remediate specific deficiencies identified in the initial placement test and to retake the test. Students self-selected to participate in the program and participation in the study was limited to students who scored below the cutscore and volunteered to review the material prior to retaking the CPT. Data pertaining to educational performance (i.e., IT, RT, RP, and GR), and demographics (i.e., A, G, and E) were extracted from the SWCC database pursuant to IRB approval.

Descriptive measures such as frequency distributions, proportions, completion rates, summary statistics, graphs and charts were used to organize, summarize, and analyze data. Following the classification of variables, univariate, bivariate and inferential statistics, such as one-sample z test for proportions, one-sample t test, two-sample t test for dependent samples, ANCOVA, and logistic regression were utilized to determine group comparability and effectiveness of independent variables (Coladarci et al., 2008; Garson, 2010; Larson & Farber, 2003; Norusis, 2006). EXCEL, StatPlus, and SAS were used to perform statistical analysis.

Collins (2008) and Kirst (2001) voiced that, in view of current trends in remedial math education at community colleges, i.e. the continually increasing number of students placed in remedial math and the inherent shortcomings in the assessment and placement process, improvements of placement processes should receive highest priority (Collins, 2008; Kirst, 2001). As many administrators at community colleges are faced with making choices to increase the effectiveness of assessment placement processes and to minimize cost to the institution and students, research was needed quickly to offer alternatives to address the remedial math dilemma at the community college level. Encouraging results of this study were not only advantageous to the institution, the student and the taxpayer, but also substantiated the critical theory approach in that the intervention allowed empowerment of students and promoted liberation (Lather, 2003). Hence, the proposed study added to the body of knowledge and possibly helped students reach higher levels of success.

Chapter 4: Findings

The purpose of this quantitative study was to explore the efficacy of implementing targeted math remediation into the assessment placement process to improve test scores, placement outcomes, and success rates of community college students. Today over 70% of students entering community college enroll in remedial math (Biswas, 2007; Brown & Niemi, 2007). Many educators, policy makers, and the public advocate a need to reduce the number of students placed in remedial math and to improve completion rates. SWCC was selected for this research due to its pioneering efforts in improving math assessment placement utilizing the newly designed criterion referenced ALEKS Math Placement Test. The instrument availed a ground-breaking design in that it not only assessed student proficiencies, but also performed an in-depth analysis of student content knowledge and provided an individualized detailed report listing topics in need of remediation (ALEKS, 2010). Furthermore, ALEKS Learning Modules, a remediation tool, was available for students to engage in review of material. In comparison to currently available computerized placement tests (CPTs), such as ACCUPLACER, COMPASS and ASSET, ALEKS was the first testing instrument utilized for assessment, review, retesting, and placement purposes.

The purpose of this chapter is to present the results of the analysis. Evaluation of results for two-tailed t-test for dependent samples, ANCOVA, one-sample z test for proportions, one-sample t test, and logistic regression for research questions posed in Chapter 1 are provided. The chapter concludes with an interpretation of findings and summary of results.

Results

Data gathered from the research site included student records and information collected from documents, campus websites, and observations. This data will seek to answer the three research questions developed for this study.

Student Records. During the pilot phase, the 2010 calendar year, 1096 new students chose to take the ALEKS Level II math placement test and 821 students (75%) scored below the cutscore set for admission to college-level math. Of the latter group, 172 students (21%) opted to engage in targeted math review and retake of CPT. This group of students constituted the sample size for this study and corresponding student records were extracted from SWCC database. Summary statistics for demographics by ethnicity and gender are provided in Table 4. The distribution of students by ethnicity closely reflected SWCC student demographics (see Table 4).

Table 4

Demographic Characteristics by Ethnicity and Gender

Characteristics	Male (%)	Female (%)	N	%	SWCC
American Indian/Alaskan Native	0 (0)	5 (6)	5	3	3%
Asian/Pacific Islander	6 (7)	8 (9)	14	8	5%
Black/African American	2 (3)	3 (4)	5	3	4%
Hispanic/Latino	17 (20)	18 (21)	35	20	17%
White	54 (64)	50 (58)	104	61	63%
Other	6 (7)	2 (2)	8	5	8%
Total	85	86	171	100	100%

Note. One student unspecified by gender in white subgroup.

An approximately equal number of females and males engaged in the intervention with 92% (159 of 171) of students age 25 years or younger. Within age groups, categories under 21 showed a larger number of females (83%) versus males (67%), while categories 21 to 25 exhibited twice the number of males compared to females (see Table 5).

Table 5

Demographic Characteristics by Age and Gender

Characteristics	Male (%)	Female (%)	N	%
Under 21	57 (67)	71 (83)	128	75
21 - 25	20 (24)	10 (12)	31	18
26 - 30	6 (7)	1 (1)	7	4
31 - 35	1 (1)	1 (1)	2	1
Over 35	1 (1)	3 (3)	4	2
Total	85	86	171	100

Note. One student unspecified by gender in age group 26-30.

Description of Research Site. The website for the research site provided information regarding the services potential students must complete for enrollment. The website detailed the steps in Admission Requirements as follows (MCC, 2010):

1. Complete a Student Information Form (SIF),
2. Apply for Financial Aid,
3. Course placement testing and academic advising,
4. Register for classes,
5. Pay tuition and fees, obtain a student ID and parking sticker.

The research site has a dedicated area for the administration of the placement test. This area is known as the Testing Services Center. Subsequent to completion of the SIF and prior to counseling, all future students are required to take the math placement test at the Testing Services Center. Only students who can provide proof of successful completion of college-level math courses attained within one year are exempt from taking the placement test. A practice exam for the placement test is available at the SWCC website (see Appendix C) and the college encourages students to practice prior to taking the placement test. According to research most students do not take the practice exam or review for the placement test (Ruiz, 2007). This finding was confirmed by SWCC testing personnel. SWCC policy does not place the burden on testing personnel to encourage students to review material prior to taking the initial placement test, but instructed testing personnel to inform students regarding their options and the differences of ALEKS I and ALEKS II level testing prior to testing. Inspite of the review option provided, almost all students take the placement test during their initial visit to the campus and prior to advising.

Upon completion of the placement test students receive a printout of test results. The results are summarized via a pie chart depicting the proportion of math topics successfully mastered (see Appendix E). The ALEKS system also creates a detailed list of math topics and subtopics in need of remediation, which is most helpful to students in reviewing material. However, to minimize expenses the college opted to only provide a printout of the summary portion as students can access the detailed progress report online (see Appendix E).

Pursuant to taking the CPT, all students receive print material explaining what to do after taking the ALEKS Math Placement Test and where to get help with review (see Appendix D). ALEKS Learning Modules is available to students to review material at a $30 fee. At any given time, 22 stations are available for testing at the testing center with an optional 25 stations available during peak time.

Subsequent to placement testing students schedule an appointment with a counselor. On average students are allotted 10 to 15 minutes for consultation. The counselor evaluates and interprets test results, and offers recommendations as to course taking and repeat placement testing. Counselors are instructed to encourage students who are not eligible to enroll in college-level math or unable to enroll in the course they wish to take to review the material and retake the CPT. Students are allowed one CPT retake. The college mandates that students enroll in the course identified via the final placement test. All remedial courses at SWCC are offered through the Math Department and the college avails three levels of developmental courses. SWCC uses successful completion of remedial coursework to assess readiness of students for college-level courses. Students are allowed two retakes per remedial course.

Research Questions. In order to explore the relationship between targeted remediation, the independent variable, and student outcomes, the dependent variables, the following three research questions and associated null and alternative hypotheses guided this research.

Q1. To what extent did targeted math remediation during the assessment placement process affect CPT scores?

$H1_0$: There was no statistically significant difference in student initial and retake CPT scores for students who engaged in targeted math remediation.

$H1_a$: There was a statistically significant difference in student initial and retake CPT scores for students who engaged in targeted math remediation.

Q2. To what extent did targeted math remediation during the assessment placement process affect student placement?

$H2_0$: There was no statistically significant difference in student placement outcomes for students who engaged in targeted math remediation.

$H2_a$: There was a statistically significant difference in student placement outcomes for students who engaged in targeted math remediation.

Q3. To what extent did targeted math remediation during the assessment placement process affect course completion rates?

$H3_0$: There was no statistically significant difference in student success rates for students who placed in college-level math as a result of the intervention compared to students who originally placed in the course.

$H3_a$: There was a statistically significant difference in student success rates for students who placed in college-level math as a result of the intervention compared to students who originally placed in the course.

Research Question Q1. The following alternative hypotheses were developed for research question one:

H1A: There was a significant difference for students who engaged in independent review in terms of mean differences in initial and retake placement test scores or $\mu_{IT} - \mu_{RT} \neq 0$.

H1B: The mean retake score was significantly greater than 37.

H1C: There was a significant difference among gender in terms of mean difference in initial and retake test scores.

H1D: There was a significant difference among ethnic groups in terms of mean difference in initial and retake test scores.

H1E: There was a significant difference among age groups in terms of mean difference in initial and retake test scores.

These alternate hypotheses were designed to test the statistical significance of the effect of the intervention on differences in IT and RT scores (H1A), on students reaching the cutscore criteria set for admission to college-level math (H1B), and the effect of gender, ethnicity, and age on retake scores (H1C, H1D, and H1E). Results aided in determining the extent to which targeted math remediation during the assessment placement process affected CPT scores.

Alternative hypothesis H1A. To determine if there was a statistically significant difference in student IT and RT scores, pretest-posttest evaluations were conducted utilizing a two-tailed paired *t* test. A sample of 172 students was included in the study. Initial (IT) and retake (RT) placement test scores measured student knowledge. Scores extracted from the college database ranged from 0 to 100. Computation of mean IT and RT scores produced scores of 20.1 (SD = 8.64) and 31.3 (SD = 16.43), respectively. Results indicated that on average students increased their score by 11.2 (SD = 12.07) points. Summary statistics for IT, RT, and differences in CPT (DIFFCPT) scores are provided in Table 6.

Table 6

Mean IT and RT Scores and DIFFCPT Scores

Characteristics	N	M	SD	Range
IT	172	20.1	8.64	2-37
RT	172	31.1	16.43	8-81
DIFFCPT	172	11.2	12.07	-9-51

To assure validity of results, the paired *t* test warranted verification of assumptions of normality and homogeneity of variances. Please note that the equality of variances assumption can be violated since the *t* test is powerful enough to overcome such limitations. The histogram for DIFFCPT scores showed an approximately normal distribution (see Figure 2). The Q-Q Plot for DIFFCPT scores depicted a trend line approximating a straight line (see Figure 3). The histogram and the Q-Q Plot verified normality of DIFFCPT test scores.

Figure 2. Histogram for DIFFCPT

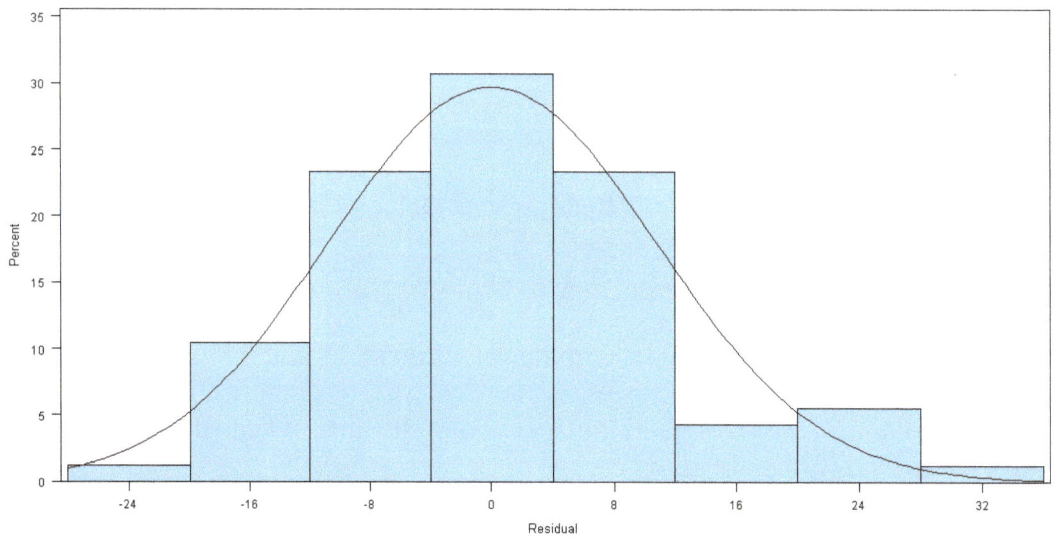

Figure 3. Q-Q Plot for DIFFCPT

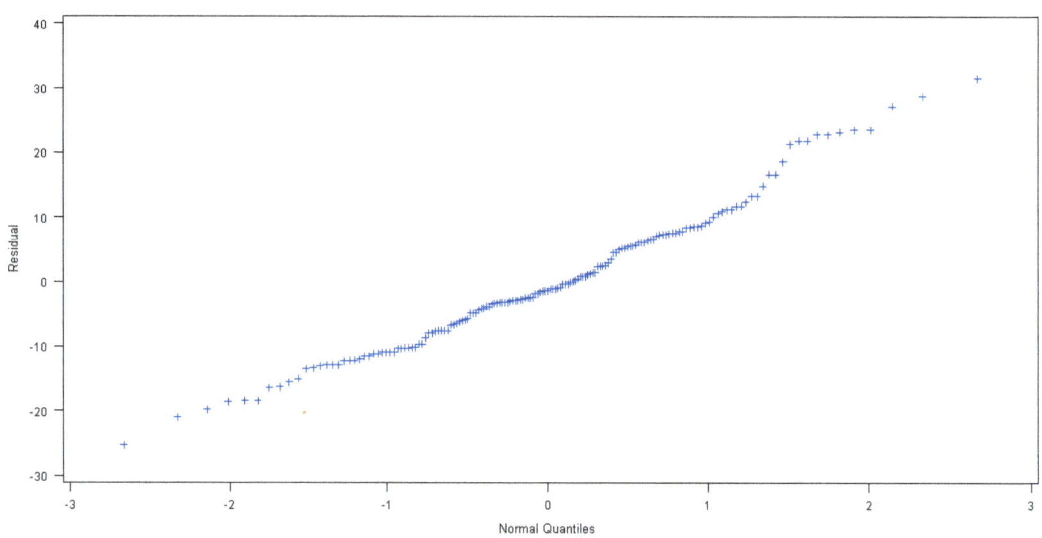

The box plot for IT and RT scores (see Figure 4) indicated differences in variability for the two groups. The *F* test for equality of variances showed a statistically significant difference in variances, $F(171, 171) = 3.62$, $p = .0000$, at $\alpha = .05$ (see Appendix F). Hence, equality of variances could not be assumed.

Figure 4. Box Plot for IT and RT Scores

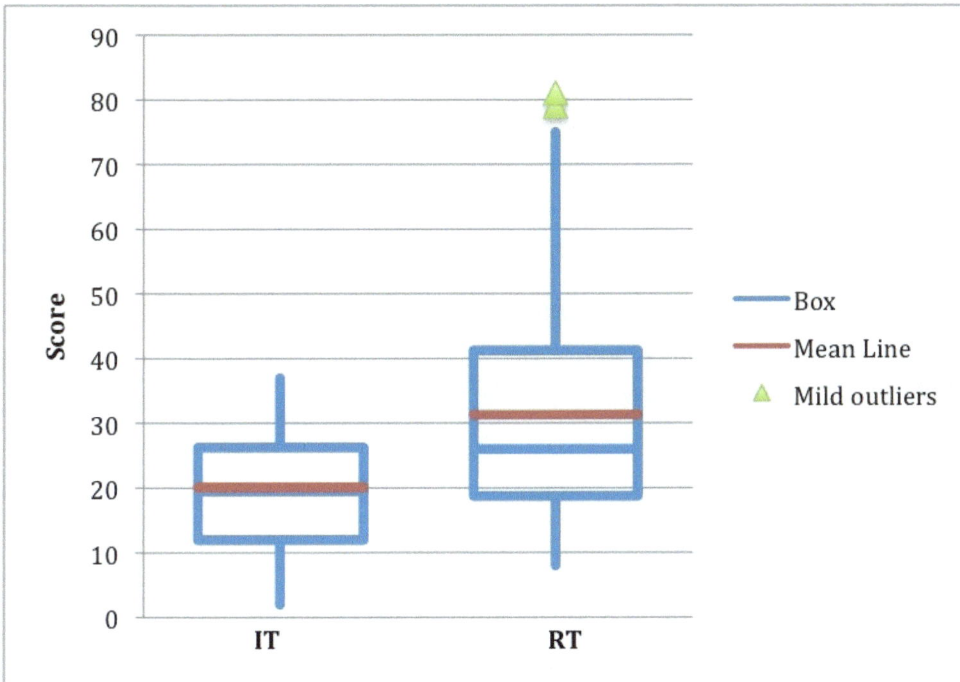

The assumptions underlying the paired *t* test substantiated, results of the paired *t* test produced enough evidence to conclude that there was a statistically significant difference in mean IT and RT CPT scores, $t(171) = 12.13$, $p < .0001$. Table 7 displays results of the paired sample *t* test. The null hypothesis was rejected at $\alpha = .05$. The 95% Confidence Interval (CI) [9.38, 13.02] for mean DIFFCPT scores excluded the hypothesized value of 0, i.e. no increase in test scores, substantiating results of the *t* test. Overall retake test scores were significantly higher for students who engaged in independent review and retake of CPT.

Table 7

Paired T Test for Differences in IT and RT Scores

The TTEST Procedure

Variable: DIFFCPT

N	Mean	Std Dev	Std Err	Minimum	Maximum
172	11.1977	12.1046	0.9230	-9.0000	51.0000

Mean	95% CL Mean		Std Dev	95% CL Std Dev	
11.1977	9.3758	13.0196	12.1046	10.9464	13.5391

DF	t Value	Pr > |t|
171	12.13	<.0001

In summary, results indicated that there was sufficient statistical evidence to conclude that students who engaged in targeted math remediation and retook the CPT improved their test scores. However, these results are not necessarily an indicator that students met the cutscore criteria. As students needed to achieve a score greater than 37 to gain access to college-level math, the statistical significance of retake scores greater than 37 was tested in hypothesis H1B.

Alternative hypothesis H1B. A one-sample *t* test was performed to determine if the mean retake score was greater than 37. Table 8 provides a summary of the number of students who met the cutscore criterion for placement in college-level math.

Table 8

RT by Cutoff Score

Characteristic		RT	
		N	%
Cutoff Score	<=37	122	71
	>37	50	29

The mean RT score for students who participated in the intervention was 31.31. Students on average improved their RT scores by 11.2 points (see Table 6) with 29% of students meeting the cutscore criteria for admission to college-level math (see Table 8). Results of one-sample *t* test revealed that there was not enough evidence to conclude that mean retake scores were statistically significantly higher than 37, $t(171) = -4.53$, $p = .99999$ (see Table 9), resulting in failure to reject the null hypothesis at $\alpha = .05$. The 95% CI [29.23, 33.39] showed limits significantly below the hypothesized value, substantiating the results of the *t* test. Hence, on average students who engaged in targeted math remediation did not meet the cutscore and did not gain access to college-level math in spite of significant increases in test scores.

Table 9

One-Sample T Test for RT

One-Sample T-Test									
	Mean	Mean LCL	Mean UCL	Standard Deviation	Sample size	Hypothesized value	Test Statistics	d.f.	p-level
Variable RT	31.31	29.23	33.39	16.48	172	37.	-4.53	171	.9999944407

Alternative hypothesis H1C, H1D, and H1E. ANOVA was performed to test the statistical significance of mean test score differences in IT and RT scores among demographic subgroups. Summary statistics of mean IT and RT scores and differences in scores by demographics are provided in Table 10.

Table 10

Summary Statistics for Mean IT and RT Scores and DIFFCPT Scores by Demographics

	Characteristics	N	%	IT	RT	Mean Diff
Gender	Female	86	50	19	27	8
	Male	85	49	21	36	15
	Unspecified	1	1	9	12	3
Age	Under 21	128	74	21	32	11
	21 - 25	31	18	18	29	11
	26 - 30	7	4	21	30	9
	31 - 35	2	1	28	45	17
	Over 40	4	2	18	20	2
Ethnicity	American Indian/Alaska Native	5	3	15	19	4
	Asian	13	8	25	48	23
	Black/African American	5	3	13	20	7
	Hispanic/Latino	35	20	19	28	9
	White	105	61	20	32	12
	Native Hawaiian/Other	1	1	29	23	-6
	Unspecified	8	5	18	21	3

The ANOVA Proc GLM mixed model was used to obtain a factorial analysis of variances to compare mean differences in CPT scores, the dependent variable, for groups defined by demographic variables gender, age and ethnicity, the independent variables, to test the non-directional alternative hypotheses H1C, H1D, and H1E at $\alpha = .05$. Covariance parameter estimates indicated heterogeneity of variances across ethnicity (see Appendix H).

Hence, an ANCOVA Proc GLM model incorporating the variable IT as a covariate was warranted to reduce variability and to increase power. Results of this analysis revealed that variables gender, $F(1,151) = 10.43, p = .0015$, and ethnicity, $F(5,151) = 15.66, p < .0001$, were statistically significant in predicting an increase in CPT scores (see Table 11), and age $F(4,151) = .49, p = .7416$, non-significant. The null hypotheses for gender, H1C, and ethnicity, H1D, were rejected. The null hypothesis for age, H1E, was not rejected. Thus, gender and ethnicity had a significant effect on student test score increases.

Table 11

ANCOVA Proc GLM Fixed Effects Estimates for Mean IT and RT Score Differences by Demographics

The Mixed Procedure

Solution for Fixed Effects

<table>
<tr><th>Effect</th><th>Ethnicity</th><th>Gender</th><th>Age_Group</th><th>Estimate</th><th>Standard Error</th><th>DF</th><th>t Value</th><th>Pr > |t|</th></tr>
<tr><td>Initial_CPT_Score__I</td><td></td><td></td><td></td><td>0.2700</td><td>0.1045</td><td>151</td><td>2.58</td><td>0.0107</td></tr>
<tr><td>Age_Group</td><td></td><td></td><td><21</td><td>5.5566</td><td>4.8689</td><td>151</td><td>1.14</td><td>0.2556</td></tr>
<tr><td>Age_Group</td><td></td><td></td><td>21-25</td><td>6.5259</td><td>5.1206</td><td>151</td><td>1.27</td><td>0.2045</td></tr>
<tr><td>Age_Group</td><td></td><td></td><td>26-30</td><td>4.6559</td><td>5.5344</td><td>151</td><td>0.84</td><td>0.4015</td></tr>
<tr><td>Age_Group</td><td></td><td></td><td>31-35</td><td>8.5377</td><td>9.5483</td><td>151</td><td>0.89</td><td>0.3727</td></tr>
<tr><td>Age_Group</td><td></td><td></td><td>>35</td><td>0</td><td>.</td><td>.</td><td>.</td><td>.</td></tr>
<tr><td>Ethnicity</td><td>American Indian/Alaska Native</td><td></td><td></td><td>14.0199</td><td>2.4782</td><td>151</td><td>5.66</td><td><.0001</td></tr>
<tr><td>Ethnicity</td><td>Asian</td><td></td><td></td><td>27.0611</td><td>4.5751</td><td>151</td><td>5.91</td><td><.0001</td></tr>
<tr><td>Ethnicity</td><td>Black/African American</td><td></td><td></td><td>15.4327</td><td>3.4014</td><td>151</td><td>4.54</td><td><.0001</td></tr>
<tr><td>Ethnicity</td><td>Hispanic/Latino</td><td></td><td></td><td>15.2221</td><td>2.4641</td><td>151</td><td>6.18</td><td><.0001</td></tr>
<tr><td>Ethnicity</td><td>White</td><td></td><td></td><td>17.0227</td><td>2.1746</td><td>151</td><td>7.83</td><td><.0001</td></tr>
<tr><td>Ethnicity</td><td>Native Hawaiian/Other</td><td></td><td></td><td>0</td><td>.</td><td>.</td><td>.</td><td>.</td></tr>
<tr><td>Gender</td><td></td><td>Female</td><td></td><td>-5.5324</td><td>1.7133</td><td>151</td><td>-3.23</td><td>0.0015</td></tr>
<tr><td>Gender</td><td></td><td>Male</td><td></td><td>0</td><td>.</td><td>.</td><td>.</td><td>.</td></tr>
</table>

Type 3 Tests of Fixed Effects

Effect	Num DF	Den DF	F Value	Pr > F
Initial_CPT_Score__I	1	151	6.68	0.0107
Age_Group	4	151	0.49	0.7416
Ethnicity	5	151	15.66	<.0001*
Gender	1	151	10.43	0.0015*

Note. * $p < .05$.

To determine subgroups of demographics, which were more likely to improve test scores, examination of least squares means estimates for age, ethnicity, and gender groups was performed and delineated the following (Table 12). Although the variable age overall proved non-significant, students in age groups <21 ($t[151] = 8.80$, $p < .0001$), and 21 to 25 ($t[151] = 5.35$, $p < .0001$), were more likely to achieve higher increases in test scores compared to other age groups and increased test scores on average by 9.2 and 10.2 points, respectively. Among ethnic groups, Asian students ($t[151] = 4.33$, $p < .0001$) showed the largest increases in test scores with an average of 20.96. Improvements achieved by Whites ($t[151] = 5.09$, $p < .0001$) and Hispanics ($t[151] = 3.85$; $p = .0002$) were also significant, but proved to be much lower with an average increase of 10.92 and 9.12, respectively. Males ($t([151] = 4.67$, $p < .0001$) were more likely than females to increase their scores and showed average increases of 11.46 points.

Table 12

Least Squares Means for GLM ANCOVA Model

Least Squares Means

Effect	Ethnicity	Gender	Age_Group	Estimate	Standard Error	DF	t Value	Pr > |t|
Age_Group			<21	9.1925	1.0450	151	8.80	<.0001*
Age_Group			21-25	10.1618	1.8977	151	5.35	<.0001*
Age_Group			26-30	8.2918	2.5606	151	3.24	0.0015
Age_Group			31-35	12.1736	8.2731	151	1.47	0.1432
Age_Group			>35	3.6359	4.9150	151	0.74	0.4606
Ethnicity	American Indian/Alaska Native			7.9179	2.4718	151	3.20	0.0017
Ethnicity	Asian			20.9592	4.8394	151	4.33	<.0001*
Ethnicity	Black/African American			9.3307	3.1861	151	2.93	0.0039
Ethnicity	Hispanic/Latino			9.1201	2.3705	151	3.85	0.0002*
Ethnicity	White			10.9207	2.1459	151	5.09	<.0001*
Ethnicity	Native Hawaiian/Other			-6.1020	2.8549	151	-2.14	0.0342
Gender		Female		5.9249	2.2023	151	2.69	0.0079
Gender		Male		11.4573	2.4523	151	4.67	<.0001*

Note. * Least squares means > 9.000 and $p < .0002$.

In summary, the findings for research question 1 indicated the following. Overall students significantly increased test scores pursuant to engaging in targeted remediation and retake of CPT. However, increases were not significant in terms of students reaching the cutscore set for admission to college-level math. Test score increases were significant among gender and ethnic groups, with male students and Asian, White, and Hispanic students more likely to significantly improve their test scores.

Research Question Q2. The following alternative hypotheses were developed for research question two:

H2A: The majority of students who engaged in independent review and retook the CPT placed in college-level math or $p_{RP} > .5$.

H2B: There was a significant difference among gender in terms of placement outcomes.

H2C: There was a significant difference among ethnic groups in terms of placement outcomes.

H2D: There was a significant difference among age groups in terms of placement outcomes.

These alternate hypotheses were designed to test the statistical significance of the effect of the intervention on placement outcomes. Specifically, to test if the majority of students improved placement from remedial math to college-level math (H2A), and if gender (H2B), ethnicity (H2C), and age (H2D) impacted student placement. Results aided in determining the extent to which the intervention affected course placement.

Analysis of data revealed that approximately 172 of 821 (21%) of students who placed in remedial math pursuant to taking the ALEKS II CPT retook the CPT. Within this group, 54% (93 students) improved their overall placement; 29% were successful in gaining access to college-level math and 25% improved placement within remedial math. Overall 46% of students did not improve placement. Regardless of placement outcomes, 43% of students did not return nor enroll in the course mandated by the college during the subsequent semester, although more than half of the students improved their overall placement (see Table 13).

Table 13

Summary of Placement Outcomes and Course Completion

Characteristic	N (%[a])	Enrolled in Mandated Course N (%[a])	Successfully Completed Course N (%[b])	Did Not Enroll in Course N (%[a])
Improved Placement from Remedial to College-Level Math	50 (29)	34	22 (65)	16
Improved Placement within Remedial Math	43 (25)	23	19 (83)	20[c]
Did Not Improve Placement	79 (46)	41	35 (71)	38[d]
Total	172 (100)	98 (57)	76 (78)	74 (43)

Note. a. Percentage with respect to total N of 172; b. Percentage with respect to total N enrolled in mandated course; c. 16 of 20 students on the edge of remediation; d. 18 of 38 students on the edge of remediation.

Alternative hypothesis H2A. A one-sample *z* test for proportions was performed to determine if the majority of students increased test scores sufficiently to gain access to college-level math (see Table 8). A total of 50 students placed in college-level math, i.e. achieved a retake score greater than 37, pursuant to the intervention. Thus, the proportion of students who improved placement to college-level math was .29 (50/172). Prior to performing the *z* test for proportions, verification of assumptions was necessary, i.e. $np \geq 5$ and $n(1 - p) \geq 5$. Computation of np and n(1- p) produced values of 25 and 25, respectively, verifying the underlying assumptions. Results of the one-sample *z* test for proportions (see Table 14) proved non-significant at $\alpha = .05$, $z = -5.49$, $p = .9999$, resulting in failure to reject the null hypothesis.

Table 14

One-Sample Z Test for Proportions for RP

PropZTest
prop>.5
z=-5.489948532
p=.9999999798
p(hat)=.2906976744
n=172

The 95% CI [.22, .36] did not include the hypothesized value underscoring the results of the *z* test. Hence, the majority, i.e. more than 50%, of students did not improve their test scores sufficiently to gain access to college-level math given the opportunity to review and retake the placement test, although students showed significant improvements in test scores.

Alternative hypothesis H2B, H2C, and H2D. Logistic regression availed a tool to test the effect of demographics on placement outcomes, i.e. test alternative hypotheses H2B, H2B, and H2C. Binary logistic regression was used to predict the binary dependent variable retake placement (RP) on the basis of continuous and categorical independent variables with age a continuous variable and gender and ethnicity categorical variables. As each student constituted a unique record, requirements of independence of variables was verified. To determine the fit of the logistic regression model, the Hosmer and Lemeshow Goodness-of-Fit Test was performed. The test produced $\chi^2(6) = 1.0053$ and $p = .9854$, indicating a good fit of data (see Table 15). Note, a p-value greater than .05 indicates a good fit.

Table 15

Goodness of Fit Test for Logistic Regression Analysis for RP

The LOGISTIC Procedure		
Hosmer and Lemeshow Goodness-of-Fit Test		
Chi-Square	DF	Pr > ChiSq
1.0053	6	0.9854

The initial logistic regression model produced questionable results necessitating redefinition of variables age to intervals <21, 21 to 30, and >30, and ethnicity to White, Hispanic, and Other (subgroup 'Other' combined all ethnic groups other than White and Hispanic). Subsequent analysis identified variables age, $\chi^2(2) = 3.7963$, $p = .1498$, and ethnicity, $\chi^2(2) = 4.1976$, $p = .1226$, non-significant and gender, $\chi^2(1) = 15.7384$, $p < .0001$, significant at $\alpha = .05$ (see Table 16). The null hypothesis for gender, H2B, was rejected. The null hypotheses for ethnicity, H2C, and age, H2D, were not rejected.

Table 16

Logistic Regression Analysis for RP by Demographics

The LOGISTIC Procedure

Model Fit Statistics

Criterion	Intercept Only	Intercept and Covariates
AIC	208.665	196.715
SC	211.807	215.565
-2 Log L	206.665	184.715

Type 3 Analysis of Effects

Effect	DF	Wald Chi-Square	Pr > ChiSq
AGECAT	2	3.7963	0.1498
RACE	2	4.1976	0.1226
Gender	1	15.7384	<.0001*

Analysis of Maximum Likelihood Estimates

Parameter		DF	Estimate	Standard Error	Wald Chi-Square	Pr > ChiSq
Intercept		1	-1.0811	0.3751	8.3058	0.0040
AGECAT	> 30	1	0.7027	0.6634	1.1221	0.2895
AGECAT	21-30	1	-0.7744	0.4392	3.1082	0.0779
RACE	HISPANIC	1	-0.7069	0.3575	3.9097	0.0480*
RACE	OTHER	1	0.3409	0.3263	1.0914	0.2962
Gender	Female	1	-0.7733	0.1949	15.7384	<.0001*

Note. * $p < .05$.

Analysis of maximum likelihood estimates presented in Table 16 showed that the odds of gender decreased slightly (-.7733) if the subject was female at $\alpha = .05$. In other words, males were more likely to improve scores than females. Although ethnicity overall proved non-significant, maximum likelihood estimates of odds for Hispanic students decreased slightly

(-.7069) at $\alpha = .05$, indicating that White students were more likely to improve test scores than Hispanics.

In summary, findings for research question 2 indicated the following. Although students significantly increased test scores pursuant to review and retake of CPT and 29% of students improved placement from remedial to college-level math, results of the one-sample *z* test showed that improvements were not substantial enough to support the claim stipulated in H2A. The majority, i.e. 50% or more, of students who engaged in the targeted math remediation did not gain access to college-level math. Overall, male and White students were more likely to improve placement than females and other ethnic groups.

Research Question Q3. The following alternative hypotheses were developed for research question three:

H3A: Students who placed in college-level math based on CPT retake were not as successful than students who originally placed in the course or $p_{SR} \neq .70$.

H3B: There was a significant difference among gender in terms of success rates.

H3C: There was a significant difference among ethnic groups in terms of success rates.

H3D: There was a significant difference among age groups in terms of success rates.

These hypotheses were designed to test the statistical significance of student success rates. Hypothesis H3A tested if students who placed in college-level math as a result of the intervention were as successful as students who originally placed in the course. Alternative hypotheses H3B, H3C, and H3D tested the effect of gender, ethnicity, and age on student success rates. Results aided in determining the extent to which targeted math remediation and CPT retake affected completion rates.

To better convey the significance of findings and the effect of the intervention on student success rates, presentation of results from summary statistics deem invaluable. The overall success rate (SR) of students at SWCC enrolled in college-level math was .70. Of the 50 students who placed in college-level math, 34 students enrolled in a college-level course during the subsequent semester with 22 successfully completing the course (see Table 13). Sixteen students postponed taking a college-level math course. The success rate of students who enrolled in a college-level math course the subsequent semester was 65%.

Success rates of students who did not place in college-level math were as follows. Approximately 25% (43 of 172 students) improved placement within remedial math. Within this group more than half of students (23 of 43) enrolled in a remedial math course the following semester with 83% successfully completing the course. Among those students who did not improve placement within remedial math (79 of 172) approximately 50% did not enroll in a remedial math course. Of those who enrolled in a remedial course, 71% successfully completed the course. Overall success rates for students in remedial math proved to be higher than those of students in college-level math.

Alternative hypothesis H3A. A two-tailed z test for proportions was used to test the hypothesis H3A. Computation of np and n(1 - p) produced values of 23.8 and 10.2, respectively, verifying assumptions underlying the z test for proportions. Results of the z test (see Table 17) showed non-significance, $z = -.6736$ and $p = .5005$, resulting in failure to reject the null hypothesis.

Table 17

One-Sample Z Test for Proportions for SR

PropZTest
prop≠.7
z=-.6736330697
p=.5005444801
p(hat)=.6470588236
n=34

The 95% CI [.49, .81] included the hypothesized value underscoring the results of the *z* test. Hence, students who placed in college-level math based on CPT retake were as successful as students who originally placed in the course.

Alternative hypothesis H3B, H3C, and H3D. To test the effect of demographics on success rates, a binary logistic model was used to predict the odds of SR on the basis of the continuous independent variable age and the categorical independent variables gender and ethnicity, i.e. test H3B, H3C, and H3D. As each student constituted a unique record, requirements of independence of variables was verified. To determine the fit of the logistic regression model, the Hosmer and Lemeshow Goodness-of-Fit Test was performed. The test produced a $\chi^2(5) = 1.3614$ and $p = .9354$ indicating a good fit of data (see Table 18). Note, a p-value > .05 indicates a good fit.

Table 18

Goodness of Fit Test for Logistic Regression Analysis for SR

The LOGISTIC Procedure		
Hosmer and Lemeshow Goodness-of-Fit Test		
Chi-Square	DF	Pr > ChiSq
1.3614	5	0.9354

The initial logistic regression model used produced questionable results necessitating redefinition of variables age to intervals <21 and >=21, and ethnicity to White, Hispanic, and Other (Category 'Other' combined all ethnic groups other than White and Hispanic). Subsequent analysis identified variables age, $\chi^2(1) = .18, p = .6680$, and ethnicity, $\chi^2(2) = 3.48, p = .1756$, non-significant and gender, $\chi^2(1) = 14.69, p = .0001$, significant at $\alpha = .05$ (see Table 19). The null hypothesis for gender, H3B, was rejected. The null hypotheses for ethnicity, H3C, and age, H3D, were not rejected.

Table 19

Logistic Regression Analysis for SR by Demographics

The LOGISTIC Procedure

Criteria for Assessing Goodness of Fit

Criterion	Value
Log Likelihood	-94.2874
Full Log Likelihood	-94.2874
AIC (smaller is better)	198.5747

LR Statistics For Type 3 Analysis

Source	DF	Chi-Square	Pr > ChiSq
AGE	1	0.18	0.6680
RACE	2	3.48	0.1756
Gender	1	14.69	0.0001*

Analysis of Maximum Likelihood Estimates

Parameter		DF	Standard Estimate	Wald Error	Chi-Square	Pr > ChiSq
Intercept		1	0.2618	1.0253	0.07	0.7984
AGE		1	-0.0204	0.0486	0.18	0.6749
RACE	HISPANIC	1	-0.8768	0.5125	2.93	0.0871
RACE	OTHER	1	0.0362	0.4578	0.01	0.9370
Gender	F	1	-1.3678	0.3733	13.42	0.0002*

Note. * $p < .05$.

Analysis of maximum likelihood estimates presented in Table 19 showed that the odds of gender decreased (-1.3678) if the subject was female at $\alpha = .05$. In other words, males were more likely to successfully complete the course than females.

In summary, the findings for research question 3 indicated the following. Students who placed in college-level math as a result of the intervention were as successful as those who originally placed in the course. Overall, male students were more likely to successfully complete the course than females.

Evaluation of Findings

Key to this study was the utilization of an effective and reliable assessment placement tool to test the efficacy of the intervention. The ALEKS CPT availed such a tool as it was the first testing instrument that allows a comprehensive approach to math assessment placement, i.e. the test was used for assessment, review, re-testing, and placement purposes. Compared to most commonly used CPTs, such as ACCUPLACER, COMPASS and ASSET, ALEKS CPT is a criterion-referenced standardized test (CRT) that identifies strengths and weaknesses of a student's math knowledge (MCCCD, 2010). The test creates an individualized detailed report for each student listing topics in need of remediation (Falmagne et al., 2003) while other CPTs merely provide a score. The CRT feature of ALEKS compares student achievement to a given standard in contrast to norm-referenced CPTs, which compare student achievement to other students and assume an underlying normal distribution of test scores, i.e. aptitude tests (Mitchell, 2006). Thus, the CRT feature makes it possible for all students to achieve the highest or lowest score (Mitchell, 2006). Over the years, application of CRTs has dramatically increased at the K-12 level (Mitchell, 2006). However, at the college-level such a movement has been missing as most colleges use norm-referenced tests with the exception of placement tests developed on a statewide or institutional basis (Mitchell, 2006).

Also key to this study was the design and implementation of the intervention strategy. The intervention was designed to help students reach higher levels of success. Current reality reflects that most entering community college students do not prepare for the test and claim they could achieve higher levels of success given a list of topics to review (Ruiz, 2007). Offering students an opportunity to review and retake the CPT availed students the opportunity for educational advancement. The goal of this research was to investigate the efficacy of the intervention on test scores, placement outcomes, and success rates and was addressed in the research posed in this study. Please note that this research was unique in that it was the first study to investigate the efficacy of ALEKS CPT in combination with the intervention strategy and the effect of demographics on student outcomes. Most prior studies utilized CPTs such as ACCUPLACER, COMPASS and ASSET in examining the effect on student outcomes and neglected to incorporate testing demographic variables. Evaluation of findings for the current study are presented below.

Research Question Q1. There was a significant increase in CPT scores for students who engaged in review and retook the test. This finding is consistent with those of Geraci (2007) and Sawyer and Welsh (as cited in Geraci, 2007) according to which retesting resulted in an increase in retake scores. The Geraci and Sawyer and Welsh studies utilized COMPASS CPT for assessment while the current study used ALEKS CPT. Geraci's study inherited significant differences in study design compared to the current study in that students were not provided a list of topics in need of review. In spite of these differences, research substantiated the positive impact of review and CPT retake on test scores.

Research Question Q2. There was no significant improvement in placement outcomes for students who engaged in the intervention. The findings of the current study contradicted with those of Ruiz (2007) and Golfin et al. (2005) who found that review courses/preparatory workshops and practice placement exams, respectively, significantly improved placement outcomes. The Ruiz and Golfin et al. studies were conducted utilizing COMPASS CPT for assessment placement in contrast to the current study which used ALEKS CPT. With respect to review of material, Golfin et al's research used a practice placement exam for review and students in Ruiz's research engaged in a broad review of topics as part of a preparatory workshop. In contrast, the current study used ALEKS Learning Modules for review providing students with a list of topics in need of remediation prior to engaging in review. Although improvements in placement outcomes did not prove to be statistically significant, 29% of students on the edge of remediation gained access to college-level math. These results coincided with those of Geraci (2007) and Sawyer and Welsh (as cited in Geraci, 2007) according to which students on the edge of remediation most likely achieved improvements. Hence, it is possible to conclude that availing students the opportunity to review prior to taking the CPT helped improve placement outcomes.

Research Question Q3. There was no significant difference in student success rates for students who placed in college-level math as a result of the intervention compared to students who originally placed in the course. These findings were similar to those of Armstrong (1999), Ruiz (2007), and Ternes (2009) who concluded that students at the edge of remediation and given the opportunity to enroll in college-level math performed as well or better than students originally placed in the course. Results of the current study contradicted those of Geraci (2007), who found that the majority of students who placed in a higher-level course pursuant to retake performed worse in subsequent courses than students who originally placed in the course. With respect to success rates by course level and ethnicity, findings contradicted those of Brown and Niemi (2007) who found that the success rate of students in college-level courses is higher than the success rate for remedial courses, but coincided in that White students are more likely to successfully complete the course than other ethnic groups (with the exception of Asian students). The current study also revealed that 58 of 122 (48%) students who did not gain access to college-level math did not enroll in a remedial course the subsequent semester (see Table 13). These findings underscored those of Biswas (2007) and Ruiz (2007), who found that students placed in a course that is merely a review of previously learned material expressed great frustration with having to repeat a course and withdrew from the course. In lieu of the results presented, it is possible to conclude that the intervention positively impacted student outcomes and helped students reach higher levels of success while possibly also alleviating student frustration and reducing dropout rates.

Summary

The purpose of this quantitative, one-group, pre-post study was to investigate the efficacy of the intervention on test scores, placement outcomes, and success rates of entering college students. Data extracted from the SWCC database were evaluated for a pilot conducted by SWCC during the calendar year 2010. Data included IT and RT CPT scores, retake placement, grades and demographics such as age, ethnicity, and gender. The results of the data analysis were presented and evaluated in this chapter.

The study showed that students who engaged in targeted review and retook the placement test significantly improved test scores, in particular the male status group and Asian, White, and Hispanic ethnic groups. Although there was no statistically significant improvement in placement outcomes, 29% of students on the edge of remediation gained access to college-level math. Completion rates of students who placed in the course pursuant to the intervention were similar to those who originally placed in the course, in particular for the male status group and the White subgroup. Overall, availing students the opportunity to improve test scores and placement outcomes helped students reach higher levels of success.

This research paralleled the concepts of Freirian research as it created opportunities for students to be liberated from possible misplacement and availed opportunities for educational advancement (Bohman, 2005). The study identified a particular problem in society, i.e. the remedial math dilemma. The intervention, the actor to change the problem, offered an avenue to reduce the number of students who were placed into developmental math courses who could be successful in college-level courses. Overall, the adaptive criterion-referenced design of the ALEKS CPT greatly improved the validity of test scores resulting in a more accurate measurement of student proficiency (Mitchell, 2006) and the intervention presented a solution to achieve a practical goal for social transformation, i.e. remedy the math dilemma. Implications, recommendation, and conclusions for this study are presented in Chapter 5.

Chapter 5: Implications, Recommendations, and Conclusions

In 2007, over 70% of students entering community college enrolled in remedial math (Biswas, 2007; Brown & Niemi, 2007). The current social reality across U.S. community college campuses indicates that the number of students placed in developmental math classes continues to increase, but research has shown that this number may be inflated (Armstrong, 2000; Ruiz, 2007). Research has shown that remedial instruction delays a student's progress toward completion of a degree, greatly reduces the probability of completing a degree, and increases cost for students, parents, institutions, and taxpayers (Brown & Niemi, 2007; Kirst & Bracco, 2004). Numerous researchers identified assessment placement at community as key contributors to this dilemma, and suggested that colleges use a new approach to math assessment placement for entering community college students, namely identifying and remediating specific deficiencies for students who had learned higher level math in their secondary education and offering students the opportunity to retake the placement exam (Abraham & Creech, 2000; Biswas, 2007; Calcagno & Long, 2008; Davis, 1999; MeKonnen & Reznichenko, 2007; Ruiz, 2007; Thompson, 2006). Currently, students who fail the CPT are not provided specific information on what to review to address their deficiencies (Kirst, 1997) and offered the opportunity to retake the CPT (Calcagno & Long, 2008; Geraci, 2007).

This quantitative study investigated the efficacy of a new intervention strategy designed to reduce math remediation rates at community colleges, and stipulated that introducing new elements into the assessment placement and remediation process, i.e. offering students the opportunity to remediate deficiencies identified in the initial CPT and retake the CPT would aid in improving student outcomes. The newly developed testing instrument ALEKS made testing of the intervention possible. SWCC, a single community college within a large multi-college district in the Southwest, was chosen for this study, as it is the first college in the nation to utilize the ALEKS system for assessment, review, retesting, and placement purposes.

Data used in the study were obtained from the SWCC database. During the 2010 calendar year a total of 172 students participated in the new intervention program and constituted the study sample. Initial and retake CPT scores were analyzed to determine if students successfully improved math competencies, placement outcomes, and success rates given the opportunity to review and retake the CPT. The primary statistical techniques employed in this study consisted of two-tailed t-test for dependent samples, ANCOVA, one-sample z test for proportions, one-sample t test, and logistic regression. Underlying assumptions for the aforementioned statistical tests were met prior to performing statistical testing.

Limitations leading up to this study included threats to internal validity such as instrument, sampling, population sample and data collection bias, and threats to external validity such as generalizability of results. The threats of instrument bias may have possibly been due to ALEKS CPT being a new commercially developed test and the limitations inherent in the program, i.e. cutscores set by the publisher. Personnel at SWCC voiced concerns about the appropriateness of current cutscores. Thus, setting of cutscores may have influenced the results of this study. Sampling of participants may have also posed limitations to the study. Math faculty at SWCC expressed concerns regarding the advising process. Faculty suspected that students may not have been properly informed as to the importance of the CPT and the high-stakes attached to the test. Faculty remarked that more students most likely would have participated in the program given more time was devoted during the counseling process to explain the impact of remedial education on college success and greater emphasis placed on encouraging students to engage in review of material and retake of CPT.

Participant selection or population bias may have also impacted internal validity of the study. For one, participants volunteered for the program and research has shown that systematic personality differences exist between volunteers and non-volunteers (Eysenck, 2004). Secondly, participants were selected based on some extreme characteristic, namely unpreparedness for college-level math. As a result, restriction of sampling range may have occurred and was taken into consideration in the data analysis process. Population sample bias posed by potential confounding variables, such as individual students' circumstances, testing conditions, motivation, medium used to review, time devoted to review, etc., may have impacted the findings of this study. These variables were not addressed in the study design and further research incorporating such variables is recommended. Other threats to validity, such as data collection, were minimized as data were screened for egregious data entry errors.

Lastly, study delimitations were inherent as the research was conducted at a single community college campus in the Southwest. However, demographics of SWCC student population make it feasible to generalize results to other community colleges across the nation as low-income and minority students are generally overrepresented at community colleges (Prince, 2005; Ruiz, 2007).

In terms of design and implementation of this study, the research conformed to current ethical guidelines. Northcentral University IRB and SWCC IRB approved the proposal for this study. Concerns with ethics for the study were minimal given all data were extracted from the SWCC database and no direct contact with human subjects was established. The anonymity of students was ensured as all personal identifiers were removed and each student record assigned a code number. Hence, informed consent and participant right to privacy were not an issue in this study. In addition, the researcher completed all required research and ethics training offered by Northcentral University.

The remainder of this chapter provides conclusions and implications for research questions posed in Chapter 1. Limitations to the study and how such affected the interpretation of the results will be discussed. The chapter will conclude with practical applications of findings, recommendations for future studies, and conclusions drawn from this study.

Implications

Following is a restatement of each of the three research questions and associated hypotheses addressed in this study including a discussion of the implications of findings for each question and limitations to the study.

Research Question Q1. To what extent did targeted math remediation during the assessment placement process affect CPT scores?

$H1_0$: There was no statistically significant difference in student initial and retake CPT scores for students who engaged in targeted math remediation.

$H1_a$: There was a statistically significant difference in student initial and retake CPT scores for students who engaged in targeted math remediation.

There was a significant increase in CPT scores for students who engaged in review and retook the test. On average students increased their score by 11 points from 20 to 31. Alternative hypotheses H1A, H1B, H1C, and H1D were developed to test the efficacy of the intervention. Implications of findings for each alternative hypothesis are provided below.

H1A: There was a significant difference for students who engaged in independent review in terms of mean differences in initial and retake placement test scores or $\mu_{IT} - \mu_{RT} \neq 0$.

The two-tailed paired t test, $t(171) = 12.13$, $p < .0001$, revealed a statistically significant difference in CPT scores at $\alpha = .05$, resulting in rejection of the null hypothesis. The 95% CI showed that the true population mean score differences will fall between 9.38 and 13.02 substantiating the results of the t test. With 150 of 172 students (87%) who engaged in the intervention improving test scores, Pearson r proved to be statistically significant, t(171)=12.8, p = .0000, indicating that 57% of improvements in test scores can be attributed to the intervention (see Appendix G).

H1B: The mean retake score was significantly greater than 37.

Results of the one-sample t test, $t(171) = -4.53$, $p = .9999$, indicated that mean retake scores were not statistically significantly higher than 37 at $\alpha = .05$. Hence, the majority of students who engaged in the intervention did not meet the cutscore criteria set for admission to college-level math.

H1C: There was a significant difference among gender in terms of mean difference in initial and retake test scores.

H1D: There was a significant difference among ethnic groups in terms of mean difference in initial and retake test scores.

H1E: There was a significant difference among age groups in terms of mean difference in initial and retake test scores.

Testing of mean test score differences for demographic subgroups via ANCOVA revealed gender and ethnicity the most significant variables in predicting differences in CPT scores. Male students significantly outperformed female students with respect to increases in retake scores. Among ethnic groups, Asian students exhibited the most significant increases in CPT scores with White and Hispanic students showing the next highest improvements.

Although the variable age was statistically non-significant in predicting CPT scores, summary statistics revealed that age groups under 21 and 21 to 25 improved test scores on average by 11 points with age group 31 to 35 showing average increases of 17 points (see Table 10).

As almost 30% of students improved their test scores sufficiently to reach the cutscore set for admission to college-level math and 54% improved overall placement (see Table 13), the results imply that a significant number of students possibly has been misclassified into remedial math courses. These findings underscored results of prior studies. A 2007 study by Ruiz concluded that current placement practices possibly inflate the number of students placed in remedial math courses. On a national scale, studies by MeKonnen and Reznichenko (2007), Brown and Niemi (2007), and Golfin et al. (2005) found that an unquantifiable number of students most likely are placed in courses below proficiency level.

Since 2007, SWCC made substantial improvements in assessment placement practices. In 2010, the college started to utilize a more effective and reliable testing instrument, namely ALEKS CPT, which provided students with a list of topics to review, and availed students the opportunity to improve test scores via the retake CPT option. The purpose of this research was to examine the efficacy of the new assessment placement process attempting to answer the following two questions: Have the new practices positively impacted test scores and student outcomes? If so, what most significantly contributed to such improvements?

Results of this research substantiated Biswas' (2007) claims that incorporating new elements into the assessment placement process, e.g. providing students with a detailed analysis of students' weaknesses and remediating such deficiencies, helps students achieve higher levels of success. Findings also underscored suggestions presented by Abraham and Creech (2000), Biswas (2007), Blair (2008), Jaschik (2008), and Thompson's (2006), who purported that availing students the opportunity to brush-up on skills better accommodates their needs, in particular students who have mastered the material previously, and resulted in test score increases. Lastly, a study by Ruiz (2007) noted that researchers have found statistically significant improvements in standardized test scores as a result of preparation workshops and coincided with results of the current study.

In summary, the efforts made by SWCC in improving the assessment placement process and availing students the opportunity to review and retake the CPT have produced very encouraging results. The intervention has helped students improve test scores and allowed over 50% of students to improve their placement (placement outcomes will be discussed in greater detail in research question 2). It is possible to conclude that the new approach to testing most likely can be credited for improvements in test scores. Further research is needed to determine if improvements in test scores can be achieved at other community colleges across the nation utilizing the new approach and medium for placement testing.

Research Question Q2. To what extent did targeted math remediation during the assessment placement process affect student placement?

$H2_0$: There was no statistically significant difference in student placement outcomes for students who engaged in targeted math remediation.

$H2_a$: There was a statistically significant difference in student placement outcomes for students who engaged in targeted math remediation.

There was no significant improvement in placement outcomes for students who engaged in the intervention. With 29% of students improving placement from remedial to college-level math, improvements proved not substantial enough to conclude that the majority of students achieved such improvements. However, data revealed that of the 172 students who engaged in the intervention 93 students (54%) improved overall placement (29% successful in gaining access to college-level math and 25% improving placement within remedial math). Alternative hypotheses H2A, H2B, H2C, and H2D were developed to test the efficacy of the intervention. Implications of findings for each alternative hypothesis are provided below.

H2A: The majority of students who engaged in independent review and retook the CPT placed in college-level math or $p_{RP} > .5$.

The one-sample *z* test for proportions, z = -5.49, p = .9999 at $\alpha = .05$, indicated that improvements in placement outcomes were not statistically significant. The majority of students did not gain access to college-level math. The 95% CI showed that the true population percentage will fall between .22 and .36, excluding the hypothesized value of .5, and underscored the results of the *z* test. However, with 93 students (54%) improving their overall placement almost 100 students were relieved of the need to take remedial math courses, and could be considered significant.

H2B: There was a significant difference among gender in terms of placement outcomes.

H2C: There was a significant difference among ethnic groups in terms of placement outcomes.

H2D: There was a significant difference among age groups in terms of placement outcomes.

Logistic regression identified variables age, $\chi^2(2) = 3.7963$, $p = .1498$, and ethnicity, $\chi^2(2) = 4.1976$, $p = .1226$, non-significant and gender, $\chi^2(1) = 15.7384$, $p < .0001$, significant at $\alpha = .05$ (see Table 16). Maximum likelihood estimates indicated that males were more likely to improve placement outcomes than females, and White students were more likely than Hispanics. Research on stereotype threat and intellectual test performance by Steele and Aronson (1995) suggested that differences in academic performance due to ethnic and biological differences may exist, but the likelihood of such explanations is greatly exaggerated. Brown and Josephs (1999) concluded that differences in performance on standardized tests may not be attributed to gender or ethnicity, but merely a result of stigmatization.

In summary, the intervention helped over 50% of students improve overall placement and 29% of students improve placement from remedial math to college-level math with male and White students more likely to improve placement, and therefore substantially reduced the number of students in need of remediation. It is possible to conclude that the new approach to assessment placement most likely can be credited for such improvements. Further research is needed to determine if improvements in placement outcomes can be achieved at other community colleges across the nation utilizing the new approach and medium for placement testing.

Research Question Q3. To what extent did targeted math remediation during the assessment placement process affect course completion rates?

$H3_0$: There was no statistically significant difference in student success rates for students who placed in college-level math as a result of the intervention compared to students who originally placed in the course.

$H3_a$: There was a statistically significant difference in student success rates for students who placed in college-level math as a result of the intervention compared to students who originally placed in the course.

Analysis of data revealed that success rates of students who improved placement to college-level math due to the intervention were similar to students who originally placed in the course with 65% of students successfully completing the course. Success rates of students in college-level math courses were lower than those in remedial math (see Table 13). Alternative hypotheses H3A, H3B, H3C, and H3D were developed to test the efficacy of the intervention. Implications of findings for each alternative hypothesis are provided below.

H3A: Students who placed in college-level math based on CPT retake were not as successful as students who originally placed in the course or $p_{SR} \neq .70$.

Results of the *z* test (see Table 4.15), $z = -.3229$ and $p = .6266$, and CI limits [.52, .83] showed that students who placed in a course as a result of the intervention were as successful as students who originally placed in the course.

H3B: There was a significant difference among gender in terms of success rates.

H3C: There was a significant difference among ethnic groups in terms of success rates.

H3D: There was a significant difference among age groups in terms of success rates.

Logistic regression identified variables age, $\chi^2(1) = .18$, $p = .6680$, and ethnicity, $\chi^2(2) = 3.48$, $p = .1756$, non-significant and gender, $\chi^2(1) = 14.69$, $p = .0001$, significant at $\alpha = .05$ (see Table 4.16). Odds ratios indicated that males were more likely to improve completion rates than females.

In summary, the intervention not only helped a significant number of students gain access to college-level math but also substantiated that students who placed in the course as a result of the intervention were as successful as those who originally placed in the course. It is possible to conclude that the new approach to assessment placement most likely can be credited for such results. Further research is needed to determine if students at other community colleges who are placed in college-level math pursuant to the intervention can achieve completion rates equal to those who originally placed in the course.

Limitations. Several factors must be considered when interpreting the results of this study. For one, a single dependent variable was used to operationalize the research questions. As illustrated in Figure 1, numerous other factors influence academic achievement. Potential confounding variables may have impacted the findings of this study, such as individual students' circumstances, motivation, medium used to review, etc. These variables were not addressed in the study and limitations due to the study design may have impacted internal validity. Further research incorporating such variables in the study design is recommended.

Secondly, Coladarci et al. (2008) noted that if participants constitute only those students who volunteered to partake in the intervention, it is possible that a true linear relationship among variables IT and RT does not exist due to a possible restriction of range. As students in this study volunteered to review and retake the CPT and all students engaged in the intervention as a result of a unique characteristic, i.e. need for math remediation, the Pearson r may reveal little about the true correlation. Careful inspection of restriction of range (see Appendix I) and testing of linearity of data via the Pearson t test, $t(171) = 12.8$, $p = .0000$ (see Appendix G), substantiated a significant linear correlation between variables IT and RT. Calculation of the coefficient of variation for IT and RT resulted in a value of $R^2 = .57$ and a Pearson product moment correlation coefficient $r = .76$ (see Appendix G). Hence, 57% of increases in test scores can be attributed to review and retake with the remaining percentage (43%) due to other factors, such as motivation, study habits, hours studied, etc. for students who participated in the intervention.

Thirdly, conducting logistic regression with other binary independent variables requires a sample size of 300 or more to achieve a medium effect size and a power of .80 at alpha = .05 (Hsieh, Bloch, & Larsen, 1998). The data extracted from the SWCC database availed a much smaller sample, i.e. 172. Hence, caution is warranted in drawing conclusions as to the significance of results.

Fourthly, the ALEKS Math Placement Test, a new commercially developed placement test, was used to assess student math proficiencies and determine test scores. The test, a standardized criterion-referenced test, is a validated instrument and has been on the market since 2009 (ALEKS, 2010). SWCC was the first community college to use this test for assessment placement and remediation purposes. Testing personnel and math faculty at SWCC anticipated that adjustments to the cutscores set by the publisher most likely will be necessary. Thus, currently used cutscores may have impacted the internal validity of results.

Fifthly, the testing instrument ALEKS, on the market for less than one year, may inherit limitations unknown to the publisher and researcher. ALEKS is the first testing instrument utilized for assessment, review, retesting and placement purposes. As the use of ALEKS for targeted review, the intervention strategy, has not been explored in any other study, this study sets precedence in examining the efficacy of a new intervention strategy in conjunction with the newly developed ALEKS CPT and Learning Modules. Research has shown that other testing instruments, such as SAT and ACT, were subject to extensive revision after years of implementation (Choroszy & Muehlberg, 2006; Mitchell, 2006). Thus, limitations inherent in the testing instrument design may have affected the results of this study and posed threats to content validity and history. Only repeat studies on the efficacy of ALEKS will establish the validity and reliability of the testing instrument.

Lastly and possibly the most significant limitation posed to this study, was inherent in the enrollment and testing process currently utilized at SWCC. This process may have impacted the sample size and composition of the sample population in terms of age, sex, and ethnicity. As noted earlier, most colleges, like their applicants, are more concerned about access to higher education (Jenkins, 2006; Kirst, 2001) and standardized tests offer the most economical choice, require little effort to implement, and are easily accessible and maintained (Ruiz, 2007). Due to the large volume of students and limited personnel at SWCC, students are unable to meet with an advisor prior to taking the CPT. While testing personnel provides information and advises students as to the difference in testing levels, the personnel does not have the time or training to provide adequate counsel to students.

The enrollment process at SWCC stipulated that students meet with a counselor for a maximum of 15 minutes pursuant to taking the CPT. Math faculty at SWCC contended that the time allotted to student counseling most likely does not afford sufficient time to properly advise students. Review of placement policy also revealed that SWCC does not use other criteria in conjunction with the placement exam (MCCCD, 2010) although research has shown that multiple assessment tools and more time spent advising and assessing student history benefit students (Armstrong, 2000; Marwick, 2004; Mekonnen & Reznichenko, 2007; Ruiz, 2007). Shortcomings in the enrollment and assessment placement process at SWCC may have possibly posed limitations to the study and impacted internal and external validity of test results.

Recommendations

Based on findings in this study, this section presents recommendations for practical application to educational practice and recommendations for future research.

Recommendations for practical applications. This research focused on the implementation of a new intervention strategy and a new placement testing instrument to reduce math remediation rates and increase student success at community colleges. Results of this study showed very encouraging results. The intervention produced significant improvements in test scores, placement outcomes and completion rates for SWCC students, and reduced the number of students placed in remedial math at SWCC. Although the college made substantial progress in improving student outcomes, examination of current enrollment and assessment placement practices at SWCC identified areas within the process to be improved. A holistic view on assessment placement and modification of certain components of the current process are highly recommended as such could significantly improve student outcomes and substantially reduce remediation rates.

According to studies by Roueche and Roueche (1999) and Boroch et al. (2010) a holistic perspective on enrollment and assessment placement serves institutions well in promoting greater student success. Roueche and Roueche (1999) suggested essential procedures and services to include: (a) mandatory orientation and counseling, (b) mandatory assessment and placement, (c) expanded pre-enrollment activities, and (d) establishment of faculty mentors (Boroch, et al. 2010). Boroch et al. (2010) recommended that strengthening college functions in the areas of advisement, assessment, coordination, and communication would produce long-term gains in student achievement. Examination of current practices and consideration of recommendations posed by Roueche and Roueche (1999) and Boroch et al. (2010) prompted the following recommendations for SWCC and community colleges in general.

For SWCC to create essential conditions that foster more effective practices at the enrollment and assessment placement level and significantly reduce the number of students placed in remedial math courses, the process should include the following elements:

(a) Mandatory orientation,

(b) mandatory pre-assessment counseling,

(c) placement testing using a criterion-referenced test and faculty mentoring,

(d) post-assessment counseling and quasi-mandatory placement, and

(f) collaboration with high schools.

The discussion that follows delineates the significance of each element proposed.

Mandatory orientation and pre-assessment counseling should take place prior to placement testing. Orientation could be incorporated in the advising session and facilitated via a prerecorded video, e.g. students watch the video at the conclusion of the advising session. Stressing the high-stakes attached to placement testing and clearly communicating the ramifications of remedial coursework should be key components of orientation and advising. Students should come away from this session with an understanding of the importance of placement testing and how such can impact their college career.

Even though today almost all colleges (92.4%) across the nation practice mandatory assessment (Collins, 2008; Gerlaugh et al., 2007; Golfin et al., 2005; Ruiz, 2007), due to the controversy surrounding such practices, caution should be used enforcing a mandatory testing and placement policy. Boroch et al. (2010) concluded that this trend may have exacerbated the math dilemma as "mandatory placement drives up the numbers of individuals who enroll in remediation without necessarily increasing their success" (Boroch et al., 2010, p. 38). Ruiz (2007) found that 70-80% of students at the edge of remediation and given the opportunity to enroll in college-level math were as successful as students who originally placed in the course.

As scarce empirical evidence exists substantiating the effectiveness of mandatory placement to improve student outcomes, a proposed quasi-mandatory placement policy may prove most effective and efficient. Quasi-mandatory placement implies a placement process that avails flexibility and consideration of individuals' unique test score circumstances, in particular for students at the edge of remediation. The proposed process stipulates that final placement of students who scored one or two points below the cutscore would be based on multiple assessment measures. Placement of students who scored below this threshold would be placed according to placement policy set by the college.

As the advantages of multiple assessment for placement have been well established in prior research (Armstrong, 2000; Boroch et al., 2010; Marwick, 2004; MeKonnen & Reznichenko, 2007; Ruiz, 2007), consideration of other assessment measures, such as ACT, SAT scores and prior math course-taking history, in addition to ALEKS CPT test scores for placement would avail insight into a student's math proficiency and help identify students who have previously mastered higher levels of math. Granting those students access to college-level math and exempting them from retake placement testing would benefit students in that it would relieve students of the need to take a remedial math course. The practice would also benefit the college in that it would lessen the volume of students in need of retaking the placement test (e.g. alleviate pressure placed on testing personnel to accommodate the large influx of students), decrease the number of students placed in remedial math, and reduce cost to the college (i.e., the college incurs a fee of $30 to $40 for each student taking the ALEKS CPT).

Availing the expertise of faculty mentors, i.e. have a member of the math faculty present at the testing center to interpret testing results, advise students on topics to review and how to best review as well as strongly encouraging students to engage in review, would help guide students in the assessment placement process. In summary, investing more time in the initial assessment placement process could produce long-term gains in student achievement and reduce cost to the student, the institution, and the taxpayer.

Lastly, Boroch et al. (2010) noted that many students are unaware of or misinformed about college readiness and placement procedures at community colleges. "The optimal time to correct misinformation is while students are still enrolled in high school" (Boroch et al., 2010, Ch. 1.1, p. 31). Bottoms and Young (2008) recommended that colleges consider assigning permanent or iterant advisors to high school campuses to explain academic standards and placement procedures at community colleges. To affect a truly comprehensive approach to assessment placement from a K-16 perspective, SWCC is encouraged to initiate dialogue with local high schools to discuss

(a) the possible assignment of college advisors facilitating college-preparatory counseling to high school seniors, and

(b) the design and implementation of review or preparatory courses for ALEKS placement testing at the high school level.

In the near future, purportedly all community colleges in the state in which SWCC is domiciled will require students to take the ALEKS CPT and use the ALEKS CPT score as the sole criteria for course placement (MCCCD, 2010). As most students currently are unaware of and unprepared for the placement test (Ruiz, 2007), the sole use of ALEKS scores most likely will exacerbate the existing remedial math dilemma. More students likely will be classified as underprepared. At the high school level emphasis is currently placed on ACT and SAT preparation. To better prepare high school students who plan to attend community college for placement testing at the community college, high schools are strongly urged to avail students ALEKS CPT preparation. Pre-enrollment programs of this nature could help create a more seamless transition between secondary and postsecondary education (Boroch et al., 2010) and most likely help more students achieve eligibility to college-level courses or improve placement within remedial math.

Recommendations for future research. Many administrators at community colleges across the nation are faced with making choices to increase the effectiveness of assessment placement processes, to reduce the number of students placed in remedial education, and to minimize cost to the institution and students. As the topic appeals to a broad audience and is of national interest more research is needed to offer alternatives to address the remedial math dilemma at the community college level. Recommendations for future research include the following.

Firstly, confounding variables, such as student motivation, individual student circumstances, time studied, medium used for review, etc., were not addressed in this study and most likely influenced the findings of this study. Examination of a comprehensive model incorporating such variables is warranted to help gain greater insight into the efficacy of the intervention.

Secondly, concerns have been voiced by math faculty at SWCC as to the appropriateness of cutscores set for the ALEKS placement test. Studies by Armstrong (1999), Hager (2004), Marwick (2004) and Ruiz (2007) have shown that 70-80% of students at the edge of remediation and placed in college-level math were as successful as those students originally placed in the course. Ternes (2009) and Ruiz (2007) found that many students who are capable of college-level math and placed in remedial math express frustration with having to repeat course material previously learned and dropped out. As little or no research has been performed to substantiate the accuracy of cutscores set by the publisher for ALEKS, research is needed to validate currently used cutscores.

Thirdly, according to a study by Gerlaugh et al. (2007) almost all colleges use mandatory assessment placement policies. Boroch et al. (2010) noted that mandatory assessment placement policies increase the number of individuals who enroll in remediation without necessarily increasing their success. As empirical evidence documenting the effects of mandated placement policies is scarce, research is needed to investigate the effect of such policies on student outcomes (Boroch et al., 2010).

Lastly, Kirst (2003), Kirst and Bracco (2004), and Ruiz (2007) noted that the remedial math dilemma and postsecondary completion problem are a result of a lack of articulated standards and clear signals concerning adequate preparation by community colleges. High school students and teachers often do not know what colleges expect in terms of meeting admission requirements (Kirst & Bracco, 2004; Ruiz, 2007). Few high schools offer review or preparatory courses for placement testing although it is customary for students to take a review course for ACT, SAT, etc. (MeKonnen & Reznichenko, 2007). A study by Kirst and Bracco (2004) has shown that, "clear, consistent, and appropriate signals and incentives improved student learning and affect students' motivation positively" (p. 4). Bridging the gap between secondary preparation, college admission, and placement testing by clearly communicating expectations and preparing students for placement testing during their senior year in high school is highly recommended. Research is needed to examine the effectiveness of pre-enrollment and advising programs at the high school level as such provide avenues to create a seamless transition between secondary and postsecondary education (Boroch et al., 2010; Bottoms & Young, 2008).

Conclusions

Today, community colleges enroll the majority of college undergraduates (Brown & Niemi, 2007). Their open-door policy and commitment to serve the local community created bridges to both higher education and career-path employment for students of diverse educational backgrounds and in particular for disadvantaged students (Boroch et al, 2010). Acknowledging the significant role of community colleges in higher education and recognizing institutional priority of community colleges to provide developmental education for all students does not diminish the fact that enrollment in developmental courses, in particular mathematics, is exacerbating at a rapid rate while the success rate of students and student progress continues to decrease at a troubling rate (Boroch et al, 2010). Studies by Biswas (2007), Brown and Niemi (2004), and numerous other researchers identified assessment placement processes at community colleges key contributors to this dilemma (Collins, 2008; Golfin et al., 2005; Kirst, 2001). A 2008 Achieving the Dream brief acknowledged the extent of the assessment placement dilemma and stressed that greater emphasis be placed on policies governing student assessment and placement in developmental education (Biswas, 2007; Collins, 2008). Due to the high-stakes nature of placement testing, improving college admission and placement practices deserves high priority (Biswas, 2007), and as Morgan and Michaelides (2005) stressed, it is imperative for the placement process to be as solid and defensible as possible.

Boroch et al. (2010) noted that, “As institutions seek to improve their efforts to provide effective developmental education” (p. 15), the most pressing questions are: “What is most important? [and] What shall we do first?” (p. 15). Roueche and Roueche (1999) remarked that, “reality is the community college ‘mission blur’, which was created by the challenge to be all things to all people” and providing educational opportunities for all students regardless of preparation (p. 35). Kouzes and Posner (2007) emphasized that it is incumbent upon community college leaders to take an initiative and become more proactive in designing effective assessment procedures for entering college students. Current events, e.g. the economic and financial crisis, coupled with the failure of currently used tools and strategies for assessment placement to produce the desired results (MeKonnen & Reznichenko, 2007), have created an urgency unlike any that community colleges have felt before (Roueche & Roueche, 1999). Community college leaders are compelled to establish remedial education policies to improve student performance of at-risk students and to minimize cost to the institution and students (Roueche & Roueche, 1999). Over the years, various efforts have been undertaken by community colleges across the nation to remedy the situation (Armstrong, 2000; Brown & Niemi, 2007; Calcagno & Long, 2008; Golfin et al., 2005; Dorrity, 2008; Hager, 2005; Nickell, 2005; MeKonnen & Reznichenko, 2007; Ruiz, 2007), but results of such efforts proved to be mixed or at best marginal.

Roueche and Roueche (1999) found that best practices in developmental education call for a holistic view on enrollment, assessment, and placement practices. The authors further noted that to achieve improvements in student success the process should include mandatory orientation, mandatory placement, and cooperation between K-12 and postsecondary institutions as well as a reliable and effective tool for assessment placement. Mandatory orientation sessions can be used to encourage entering students to address remediation needs (Boroch et al., 2010). Research by Boylan and Saxon (2001) found that students who participate in student orientations are more likely to be retained in community college compared to students who did not receive orientation.

Even though today almost all colleges (92.4%) across the nation practice mandatory assessment (Collins, 2008; Gerlaugh et al., 2007; Golfin et al., 2005; Ruiz, 2007), due to the controversy surrounding mandatory testing and placement policy caution should be used enforcing such policies. A more individualized approach, i.e. a proposed semi-mandatory placement, deems more appropriate and most likely would result in more accurate placement of students, in particular those on the edge of remediation.

According to Roueche and Roueche (1999), most important in affecting positive student outcomes is the creation of a more seamless web of collaboration with other educational institutions, e.g. for colleges to develop and implement partnerships with public schools, has the greatest potential for achieving college readiness for first-time students. The authors further remarked engaging in the blame game as to the causes of poor educational performance in higher education requires too much time and energy. Such practices should be put aside in the interest of using time more wisely (Roueche & Roueche, 1999).

The initiative taken by SWCC to improve student outcomes and examined in this study used a more comprehensive approach to math assessment placement as well as a new groundbreaking and more effective testing technology. The intervention was unique in that it used ALEKS CPT and Learning Modules, which lent itself for assessment, review, retesting, and placement purposes and was the first testing instrument to (a) utilize criterion-referenced testing, (b) incorporate a sufficient number of questions, (c) perform an in-depth analysis of student competencies and provide a detailed list of topics in need of remediation, and (d) avail a tool for review and remediation.

The new approach to assessment placement by SWCC produced very encouraging results. The college successfully reduced the number of students placed in remedial math, improved placement outcomes, and course completion rates. The study substantiated the critical theory approach in that the intervention created opportunities for students to be liberated from possible misplacement and availed opportunities for educational advancement. In this case, the intervention helped groups or individuals that may have been marginalized, i.e. students who were capable of success in college-level math but placed into remedial math. The study added to the body of knowledge as it helped students reach higher levels of success. Limitations of this study and shortcomings inherent in the assessment placement process at SWCC warrant further research to substantiate the efficacy of this new approach and the reliability and validity of the new assessment placement tool ALEKS.

Boroch et al. (2010) stressed that a holistic view on remedial education can pay big dividends to students and the institution. Improving assessment placement processes and policies and creating a seamless transition between secondary and postsecondary education would help improve preparation for college, reduce the need for postsecondary remediation, reduce time-to-degree and increase college completion rates as well as reduce or at least contain educational expenditures (Delta Project, 2008; Kirst, 2001). In a larger context, the benefits could be substantial for all parties involved, i.e. students, parents, taxpayers and the nation, both from a socio-economic standpoint and perhaps long-term in that is would help our workforce compete in this global economy (Delta Project, 2008; National Center, 2008).

References

Abraham, A. A., & Creech, J. D. (2000). *Reducing remedial education: What progress are states making?* Atlanta, GA: Southern Regional Education Board. Retrieved from http://www.eric.ed.gov/

ACT. (1993, May). *Incremental validity of ACT assessment scores and high school course information for freshman course placement*. Iowa City, IA: ACT Research Report Series. Retrieved from http://www.eric.ed.gov/

ACT. (2003). *Eighth grade students underestimate college preparation needs*. Retrieved from http://www.act.org/news/releases/2003/7-02-03.html

ACT. (2009). *America's most widely accepted college entrance exam*. Retrieved from http://www.act.org/aap/

ACT. (2010). *ACT: Resources for education and workplace success*. Iowa City, IA: ACT, Inc. Retrieved from http://www.act.org

ALEKS. (2010). *Overview of ALEKS*. UC Regents and ALEKS Corporation. Retrieved from http://www.aleks.com

Armstrong, J. (1991, November 15). *Assessing skills from placement to co*mpletion. Paper presented at the North Texas Community College Conference, Denton, TX. Retrieved from http://www.eric.ed.gov/

Armstrong, W. B. (1994). *Math placement validation study: A summary of the criterion related validity evidence and multiple measures data for the San Diego Community College District.* San Diego, CA: Office of Institutional Research, San Diego Community College District. Retrieved from http://www.eric.ed.gov/

Armstrong, W.B. (1999). *Explaining community college outcomes by analyzing student data and instructor effects* (Unpublished doctoral dissertation). University of California, Los Angeles. Retrieved from http://www.eric.ed.gov/

Armstrong, W. B. (2000). The association among student success in courses, placement testing scores, student background data, and instructor grading practices. *Community College Journal of Research and Practice, 24*(8), 681-695. Retrieved from http://www.eric.ed.gov/

Astin, A. (2002). *Assessment for excellence: The philosophy and practice of assessment in higher education.* Westport, CT: American Council on Education Oryx Press.

Biswas, R. R. (2007). *Accelerating remedial math education: How institutional innovation and state policy interact.* Boston, MA: Achieving the Dream/Jobs for the Future. Retrieved from http://hdl.handle.net/10207/8673

Blair, R. (2008, October). Enrollments in developmental mathematics are increasing! What should be done? *AMATYC News, 23(*5).

Blair, R. (2009, March). Is developmental algebra working? *AMATYC News, 24(*2).

Bohman, J. (2005, March 8). Critical theory. In *Stanford encyclopedia of philosophy*. Retrieved from http://plato.stanford.edu/entries/critical-theory/#3

Boroch, D., Hope., L., Smith, B., Gabriner, R., Mery, P., Johnstone, R., & Asera, R. (2010). *Student success in community colleges: A practical guide to developmental education.* San Fancisco, CA: Jossey-Bass. Retrieved from http://books.google.com

Bottoms, G., & Young, M. (2008). *Lost in transition: Building a better path from school to college and careers.* Southern Regional Education Board. Retrieved from http://www.eric.ed.gov/

Boylan & Saxon. (2001, Winter). The cost of remedial education in higher education. *Journal of Developmental Education, 25*(2), 2-8. Retrieved from ProQuest Database (EJ641630).

Brown, R. S., & Niemi, D. N. (2007). *Investigating the alignment of high school and community college assessments in California*. San Jose, CA: National Center for Public Policy and Higher Education. Retrieved from http://www.highereducation.org/reports/brown_niemi/index.shtml

Brown, R. P., & Josephs, R. A. (1999). A burden of proof: Stereotype relevance and gender differences in math performance. *Journal of Personality and Social Psychology, 76*(2), 246-257. Retrieved from http://www.imamu.edu.sa/

Calcagno, J. C., & Long, B. T. (2008). *The impact of postsecondary remediation using a regression discontinuity approach: Addressing endogenous sorting and noncompliance*. Cambridge, MA: National Bureau of Economic Research. Retrieved from http://www.nber.org/tmp/13281-w14194.pdf

Choroszy, M., & Muehlberg, J. (2006, Fall). The impact of the revamped SAT on admissions policies among western land grant institutions. *College and University, 81*(2), 51-52. Retrieved from ProQuest database.

Coladarci, T., Cobb, C., Minium, E., & Clarke, R. (2008). *Fundamentals of statistical reasoning in education*. Hoboken, NJ: John Wiley & Sons.

College Board. (2009a). *Advising & placement*. Retrieved from http://professionals.collegeboard.com/higher-ed/placement/

College Board. (2009b). *ACCUPLACER*. Retrieved from

http://professionals.collegeboard.com/higher-ed/placement/accuplacer/

College Board. (2009c). *CLEP*. Retrieved from http://professionals.collegeboard.com /higher-ed/placement/clep/

Collins, (2008). *It's not about the cut score: Redesigning placement assessment policy to improve student success.* Boston, MA: Achieving the Dream/Jobs for the Future. Retrieved from http://www.achievingthedream.org/_pdfs/

Cozby, P. (2009). *Methods in behavioral research.* New York, NY: McGraw-Hill.

CSU. (2009). *CSU transforming course design for developmental mathematics*. Retrieved from http://groups.google.com/group/csu-transform-dev-math-teams/web

Davis, J. L. (1999, Fall). Developmental education paradox. *Update on Research and Leadership, 10*(3). Champaign, IL: Office of Community College Research and Leadership, University of Illinois at Urbana-Champaign. Retrieved from http://occrl.ed.uiuc.edu/Newsletter/1999/fall1999_7.asp

Delta Project. (2008). *The growing imbalance: Recent trends in U.S. postsecondary education finance.* Washington, DC: Delta Project on Postsecondary Education Costs, Productivity and Accountability. Retrieved from http://www.deltacostproject.org/resources/pdf/imbalance20080423.pdf

Dorrity, W. (2008). *Predicting student performance in developmental and college level course work in the community college* (Doctoral dissertation, Northcentral University). Retrieved from ProQuest Dissertations and Theses database. (UMI No. 3329380)

Education Trust. (1999, Fall). Ticket to nowhere. *Thinking K-16, 3*(2), 16-31. Retrieved from http://www2.edtrust.org/NR/rdonlyres /1196FBF0-FB01-4B75-B363-B1D525869F29/0/k16_fall99.pdf

Edwards, V., Olson, L., & Swanson, C. (2007, January). From cradle to career: Connecting American education from birth through adulthood. *Education Week, 26(17).* Retrieved from http://www.edweek.org/ew/toc/2007/01/04/index.html

Eysenck, W. (2004). *Research methods: Design of investigations.* New York, NY: Psychology Press. Retrieved from http://ltelelearning.org

Falmagne, J-C, Cosyn, E, Doignon, J-P, & Thiery, N. (2003). *The assessment of knowledge, in theory and in practice*. Irvine, CA: Institute for Mathematical Behavioral Sciences. Retrieved from http://escholarship.org/us/item/1bz9p41c/

Felder, J., Finney, J, & Kirst, M. (2007). *"Informed self-placement" at American River College: A case study*. San Jose, CA: The National Center for Public Policy and

Higher Education. Retrieved from ProQuest database.

Foley-Peres, K., & Poirier, D. (2008). College math assessment: SAT scores vs. college math placement scores. *Educational Research Quarterly, 32*(2), 41-49. Retrieved from ProQuest database.

Garson, D. (2010). *Logistic regression.* Retrieved from http://faculty.chass.ncsu.edu /garson/PA765/logistic.htm

Geraci. S. (2007). *Effective mathematics placement testing strategies: a study of mathematics placement test retake policy at a two-year public community college in Florida* (Doctoral dissertation, George Mason University). Retrieved from ProQuest Dissertations and Theses database. (UMI No. 3294021)

Gerlaugh, K., Thompson, L., Boylan, H., & Davis, H. (2007). National study of developmental education II: Baseline data for community colleges. *Research in Developmental Education, 20*(4). Retrieved from http://www.ncde.appstate.edu /resources/reports/documents/RiDE%2020-4.pdf

Golfin, P., Jordan, W., Hull, D., & Monya, R. (2005*). Strengthening mathematics skills at the postsecondary level: Literature review and analys*is. Washington, DC: U.S. Department of Education, Office of Vocational and Adult Education. Retrieved from http://www.cna.org/documents/IPR11659.pdf

Hager, M. J. (2005). *Avoiding the false negative: Placing students into mathematics courses according to their abilities* (Doctoral dissertation). Retrieved from http://www.ohiolink.edu/handle/2374.OX/9418/

Houser, J. (2007, March). How many are enough? Statistical power analysis and sample size estimation in clinical research. *Journal of Clinical Research Best Practices, 3*(3). Retrieved from http://learners.ncu.edu/syllabus /download_file.asp?syllabus_rr_id=136863/

Hsieh, F. Y., Bloch, D. A., & Larsen, M. D. (1998). A simple method of sample size calculation for linear and logistic regression. *Statistics In Medicine, 17*, 1623-1634. Retrieved from http://personal.health.usf.edu/ywu/logistic.pdf

Indiana Commission for Higher Education. (2008). *Reaching higher with college completion.* Retrieved from http://www.che.in.gov

Jaschik, S. (2008, July 31). *Questioning the value of remedial education.* Retrieved from http://www.insidehighered.com/layout/set/print/news/2008/07/31/remedial/

Jenkins, D. (2006). *What community college management practices are effective in promoting student success? A study of high- and low-impact institutions.* Retrieved from http://www.eric.ed.gov/

Jungnickel, P. W., & Creswell, J. W. (1994, April). Workplace correlates and scholarly performance of pharmacy clinical faculty members. *Research in Higher Education, 35*(2), 167-194. Retrieved from http://www.jstor.org/pss/40196085/

Kemmis, S. (2006, December). Participatory action research and the public sphere. *Educational Action Research, 14*(4), 459-476. doi: 10.1080/09650790600975593

Kirst, M. (1997). Exam confusion: Admissions and placement tests lack standardization. *Crosstalk*, Win 3. San Jose, CA: National Center for Public Policy and Higher Education. Retrieved from http://www.stanford.edu/group/bridgeproject/confusion/confusion.html

Kirst, M. (1998, September). Bridging the remediation gap. *Education Week, 18*(1), 76-78. Retrieved http://www.stanford.edu/group/bridgeproject/gap/remedgap.pdf

Kirst, M. (2001). *Overcoming the high school senior slump: New education policies. Perspectives in Public Policy: Connecting Higher Education and the Public Schools.* Washington, DC: Institute for Educational Leadership. Retrieved from http://www.eric.ed.gov/

Kirst, M. (2003). *College preparation and grade 12 NAEP.* Stanford, CA: National Assessment Governing Board. Retrieved from www.stanford.edu/ group /bridgeproject/College%20Prep%20and%20Grade%2012%20NAEP.pdf

Kirst, M., & Bracco, K. R. (2004). Bridging the great divide: How the K-12 and postsecondary split hurts students, and what can be done about it. In M.E. Kirst & A. Venezia (Eds.), *From high school to college: Improving opportunities for success in postsecondary education* (pp. 1-30). San Francisco, CA: Jossey-Bass. Retrieved from http://books.google.com

Kirst, M. W., & Venezia, A. (2006, March 10). What states must do. *The Chronicle of Higher Education, 52*(27), B.36. Retrieved from http://chronicle.com

Knowlton, E. (2008). *Revamping remedial math education at community colleges.* Retrieved from http://www.adventures-in-education.com /2008%20Survey%20Results.htm

Kouzes, J. M., & Posner, B. Z. (2007). *The leadership challenge*. San Francisco: Jossey-Bass.

Larson, R., & Farber, B. (2003). *Elementary statistics: Picturing the world.* Upper Saddle River, NJ: Prentice Hall.

Lather, P. (2003). Issues of validity in openly ideological research: Between a rock and a soft place. In Y. Lincoln & N. K. Denzin (Eds.), *Turning points in qualitative research: Tying knots in a handkerchief* (pp. 185-216). Walnut Creek, CA: Altamira Press. Retrieved from http://books.google.com

Levin, H. M., & Calcagno, J. C. (2008, January). Remediation in the community college: An evaluator's perspective. *Community College Review, 35*(3), 181-208. Retrieved from http://www.communitycollegereview.com

Marwick, J. D. (2004). Charting a path to success: The association between institutional placement policies and the academic success of Latino students. *Community College Journal of Research and Practice, 28*, 263-280. Retrieved from www.eric.ed.gov/

Marzano, R., & Kendall J. (2007). *The new taxonomy of educational objectives.* Thousand Oaks, CA: Sage Publications Company. Retrieved from http://books.google.com

MCC. (2010). *Recruitment Services.* Retrieved from http://www.mesacc.edu /students/recruitment/

MCCCD. (2010). *MCCCD ALEKS math placement test*. Maricopa Community College District. Retrieved from http://www.mc.maricopa.edu

Mekonnen, A., & Reznichenko, N. (2007, February). *Assessment of computerized placement test for college mathematics at Baltimore City Community College*. Paper presented at the Annual Meeting of the Eastern Educational Research Association (EERA), Clearwater, FL. Retrieved from http://www.eric.ed.gov/

Mitchell, R. (2006). *The nature of assessment: A guide to standardized t*esting. Center for Public Education. Retrieved from http://www.centerforpubliceducation.org /site/c.kjJXJ5MPlwE/b.1501925/k.C980 /The_nature_of_assessment_A_guide_to_standardized_testing.htm

Morgan, D. L., & Michaelides, M. P. (2005). *Setting cut scores for college placement* (Report No. 2005-9). New York, NY: The College Board. Retrieved from http://professionals.collegeboard.com/profdownload /accuplacer-setting-cut-scores.pdf

National Center. (2008). *Measuring up 2008*. San Jose, CA: National Center for Public Policy and Higher Education. Retrieved from http://measuringup2008.highereducation.org

National Center for Education Statistics. (2003). *Remedial education at degree-granting postsecondary institutions, Fall, 2000*. Washington, DC: U.S. Department of Education, Institute of Education Sciences. Retrieved from http://nces.ed.gov /pubs2004/2004010.pdf

National Center for Public Policy and Higher Education. (2008, March). Policy alert.

Retrieved from http://www.highereducation.org/reports /pa_mixed_signals/the_study.shtml

NCTM. (2009). *Economic impact of the achievement gap.* Webposting retrieved from http://www.nctm.org/news/content.aspx?id=%2023086#3/

Nickell, M. (2005). *The effects of developmental algebra on community college students* (Doctoral dissertation, Tennessee State University). Retrieved from ProQuest Dissertations and Theses database. (UMI No. 3206710)

Norusis, M. (2006). *SPSS 14.0 guide to data analysis.* Upper Saddle River, NJ: Prentice Hall.

Parmer, P., & Cutler, J. (2007). Easing the transition: Building better bridges between developmental and college-level math. *Journal of Applied Research in the Community College, 15*(1), 37-46. Retrieved from http://www.aypf.org /forumbriefs/2008/fb061308.htm

Parr, B., Edwards, C., & Leising, J. (2006). Effects of math-enhanced curriculum and instructional approach on the mathematics achievement of agricultural power and technology students: An experimental study. *Journal of Agricultural Education 47*(3), 81-84. Retrieved from EBSCOhost database.

Pavelcheck, D., Stern, P., & Olson, D. (2002). *Relationship between the WASL and placement tests used by community and technical colleges.* Olympia, WA: SESRC-Puget Sound Division. Retrieved from http://www.sbctc.ctc.edu/ewag /ewag0203/2002%20March%20WASL%20Placement%20STudy%20v3.doc

Prince, H. (2005). *Standardization vs flexibility: State policy options on placement testing for developmental education in community colleges.* Boston, MA: Achieving the Dream. Retrieved http://www.achievingthedream.org/_images /_index03/Policy_brief-StandardFlex.pdf

Roth, J., Crans, G., Carter, R., Ariet, M., & Kernick, M. (2001). Effect of high school course-taking and grades on passing a college placement test. *The High School Journal, 84*(2), 72-87. Retrieved from http://www.eric.ed.gov/

Roueche, J., & Roueche, S. (1999). *High stakes, high performance: Making remedial education work.* Washington, DC: Community College Press. Retrieved from http://books.google.com

Ruiz, T. (2007). *Mathematics placement tests in the community college: Possible interventions to reduce false-negative placement* (Doctoral dissertation, Arizona State University). Retrieved from ProQuest Dissertations and Theses database. (UMI No. 3243894)

Shults, C. (2000). *Remedial education: Practices and policies in community colleges.* Washington, DC: American Association of Community Colleges. Retrieved from http://www.eric.ed.gov/

Shuttleworth, M. (2008). *Quasi-experimental design.* Retrieved from http://www.experiment-resources.com

Siniscalco, M., & Auriat, N. (2005). *Questionnaire design.* International Institute for Educational Planning/UNESCO. Retrieved from http://www.scribd.com/doc /6083257/Questionnaire-design/

Smydo, J. (2008, December 1). Remedial courses used by many to adjust to college. *Pittsburgh Post-Gazette.* Retrieved from www.post-gazette.com/pg/ 08245 /908603-52.stm

Steele, C. M., & Aronson, J. (1995). Stereotype threat and the intellectual test performance of African Americans. *Journal of Personality and Social Psychology, 69*, 797-811. Retrieved from http://www.ncbi.nlm.nih.gov/pubmed/7473032

Tashakkori, A., & Creswell, J. W. (2007). Exploring the nature of research questions in mixed methods research. *Journal of Mixed Methods Research, 1*, 207. doi: 10.1177/1558689807302814

Taylor-Powell, E. (1998, May). *Questionnaire design: Asking questions with a purpose.* The Texas A&M University System. Retrieved from http://learningstore.uwex.edu/assets/pdfs/G3658-2.PDF

Ternes, R. (2009, April). *A patterned world: The shape of college placement instruments: Exploring the fundamental relationships between assessment and outcomes* [Web log message]. Retrieved from http://www.apatternedworld.com/2009/04 /shape-of-college-placement-instruments.html

Thompson, L. (2006, February 1). "Math gap" equals trouble for students heading for college. *The Seattle Times.* Retrieved from http://seattletimes.nwsource.com /html/home/

Trochim, W., & Donnelly, J. (2007). *The research methods knowledge base* (3rd ed.). Mason, OH: Thomson. Retrieved from www.stanford.edu/group /bridgeproject/PACE/Chapter8_a.pdf

Venezia, A. (2000). *Connecting California's K-12 and Higher Education systems: Challenges and Opportunities.* Stanford, CA: Stanford University. Retrieved from www.stanford.edu/group/bridgeproject/PACE/Chapter8_a.pdf

Venezia, A., Kirst, M. W., & Antonio, A. L. (2003, May). Fix K-16 disconnections, or betray the college dream. *The Education Digest, 68*(9), 34-40. Retrieved from http://www.eddigest.com

Appendixes

Appendix A:

Distribution of Topics on Standardized Math Tests

Percentage of questions devoted to:						
	Algebra 1	*Geometry*	*Data, probability, statistics*	*Number theory, arithmetic, combinatorics, logic*	*Algebra 2*	*Trigonometry/ precalculus*
Privately developed high school assessment tests						
TerraNova	14	29	23	21	0	0
Stanford 9 m/c	29	25	25	21	0	0
State high school assessment tests						
Kentucky (CATS)	9	33	17	18	20	0
Massachusetts (MCAS 10)	23	28	13	18	13	5
New York	29	26	9	26	9	3
Texas (TAAS)	12	23	3	53	0	0
College admissions exams						
SAT 1	47	23	3	23	3	0
ACT	25	27	5	18	12	8
Privately developed college placement tests						
Compass	14	23	0	19	25	15
Accuplacer al	25	0	0	0	75	0
Accuplacer cl	16	0	0	0	63	21

* al = Algebra
**cl = Calculus

Source: Education Trust, 1999.

Appendix B:

ALEKS Math Course Placement Chart

ALEKS placement scores are valid for one year prior to the first day of the semester.

Scores from the ALEKS Level I Math Placement Test

0-32	Basic Arithmetic (MAT081, 082, 083)
33-73	Introductory Algebra (MAT090, 091, 092) or Mathematical Concepts/Applications (MAT102)
74-87	Intermediate Algebra (MAT120, 121, 122)
88-100	Intermediate Algebra (MAT120, 121, 122) or take ALEKS Level II Math Placement Exam

Scores from the ALEKS Level II Math Placement Test

0-13	Take ALEKS Level I Math Placement Test or Basic Arithmetic (MAT081, 082, 083)
14-37	Intermediate Algebra (MAT120, 121, 122)
38-46	College Mathematics (MAT140, 141, 142) or College Algebra/Functions (MAT150, 151, 152)
47-63	Math for Elementary Ed (MAT156), Finite Math (MAT172), Trigonometry (MAT182), Precalculus* (MAT187), Statistics (MAT206), or Brief Calculus (MAT 212)
64-100	Calculus with Analytic Geometry I (MAT220, 221)

* Precalculus (MAT187) is a very accelerated course that combines College Algebra and Trigonometry. It is strongly recommended that students who enroll have some prior knowledge of trigonometry.

MCCCD Math Course Numbers: The number of course credits earned for a particular math course may vary even though the course content does not. This is indicated by the last digit of the course number. The table below explains the differences between the courses.

Course number ends in:	Target Population
0	This is a 5 credit course which equates to 5 academic hours of instruction per week. It is designed for students who will benefit from a slower learning pace and additional time on task with an instructor.
1	This is a 4 credit course which equates to 4 academic hours of instruction per week. The content is the same, but the pace is more accelerated than the course ending with a 0.
2	This is a 3 credit course which equates to 3 academic hours of instruction per week. This is the most accelerated course designed for students with very strong prerequisite skills.

Revised 1-Sept-09

Source: MCCCD, 2010.

Appendix C:

Placement Test Sample Questions Information

MESA COMMUNITY COLLEGE

PLACEMENT TEST SAMPLE QUESTIONS ~~

- *are available at our website:*

www.mc.maricopa.edu/testing

click on **"Sample Test Questions"** (at left under "Student Services")

Aleks

Your Success is Our Success!

Rev.. 11/2009

TESTING CENTER HOURS

Monday – Thursday 8:00am – 7:00pm (no test given out after 6:00)
Friday 8:00am – 5:00pm (no test given out after 4:00)

Closed on FRIDAYS during summer

All placement testing for English, Reading, and Math is computer-based with no time limit. Allow enough time to complete all testing. Testing schedule hours may change during peak registration times.

Note: **New students should allow approximately 2-1/2 hours to complete all three placement tests.**

Our testing services also include:

- CELSA (English as a second language)
- Instructors' make-up exams
- GED testing (Southern & Dobson, only) **M-Th by appt.**
- Nursing entrance test (Red Mountain, only) **480-654-7745**

PHOTO I.D. REQUIRED FOR ALL TESTING

Southern & Dobson: 480-461-7336 *Red Mountain : 480-654-7600*

Rev. 11/2009

Source: MCCCD, 2010.

Appendix D:

ALEKS Math Placement Test Information

Mesa Community College
Math Placement Test Information

Revised 8/19/09

What to do **BEFORE** taking the
ALEKS Math Placement Test

Please note: *If you do not have a high school diploma and have not passed the GED then this is NOT the test for you. Please ask the Testing Center staff to help you determine the placement test that best meets your needs.*

The ALEKS Math Placement Test uses test-adaptive questions; it is not a multiple-choice exam. Initially, the questions will be drawn from several levels of math and will vary in difficulty. As the exam continues, the system will gradually focus questions to your specific knowledge. **Once you have answered a question, you will not be able to go back and change your answer.**

The length of the test is between 25 and 35 questions. The exact number will vary due to the nature of the test. It is likely you will be asked some questions on material that you do not know. In that case, answer "I do not know". However, if you believe you may be able to answer the question, it is best to try. ALEKS interprets "I do not know" as meaning that you have never seen the information before and could result in an under-evaluation of what you know. **If you do not have enough time to finish the test you can log out and come back another time to finish.** Additional information can be found on this website: **http://www.mesaccc.edu/mathplacement**

Two separate exams are used to accurately measure your current mathematical knowledge and to provide appropriate review topics. **ALEKS Level I Math Test** places you into Basic Arithmetic, Introductory Algebra, and Intermediate Algebra. **ALEKS Level II Math Test** places you into Intermediate Algebra, College Mathematics, College Algebra/Functions, Finite Math, Trigonometry, Brief Calculus, and Calculus I.

How do I know which test to take?
You should take the **ALEKS Level II Math Test** if ALL of the following apply:

- You received a C or better in high school advanced math, pre-calculus, or college algebra AND
- You completed the course within the past 12 months, AND
- You are confident that you have retained your knowledge of algebra concepts and skills. For example, you can correctly solve quadratic, rational, and radical equations.

Otherwise, take the **ALEKS Level I Math Test**

When you take the ALEKS test, you will establish a log-in name and password. Record and keep this information for future use:

Login Name: ______________________________

Password: ______________________________

Mesa Community College Revised 8/19/09
Math Placement Test Information

What to do **AFTER** taking the
ALEKS Math Placement Test

Q: How many times can I take the test?
A: You may take two ALEKS Math tests per year: either two of the same level or one of each level

Q: How long is the test score valid?
A: The ALEKS Math Placement Test score is valid for one year and must also be valid on the first day of the semester

Q: What if I do not obtain the required minimum score for the math course I want to take?
A:

- Enroll in the math course you placed into, or
- You may take the ALEKS test one time in an attempt to improve your score. You are urged to use the review topics that are part of your ALEKS test results to guide your review before retaking the test.

Q: Where can I get help with my review?
A: The ALEKS learning system has online tutorials for purchase ($30.00 for 6 weeks) that are aligned with the math course you would like to take. Or you can use a textbook from a previous math class to study the topics suggested by your ALEKS test results. The MCC Math Learning Center can also be of help. Information on the ALEKS system tutorials and other review options can be found on this website:
http://www.mc.maricopa.edu/dept/d25/placement/math_placement_review.html

Q: Where can I get math advising support?
A: Contact the MCC Math Learning Center in the Math Department or the Learning Enhancement Center in the Library. Additional information can be found on this website:
http://www.mc.maricopa.edu/dept/d25/placement/math_advisement.html

Q: Will I be able to access my test results after I leave the Testing Center?
A: Yes, you can see your scores by logging onto **http://www.ALEKS.com/maricopa**
Use your login name and access code established when you began the ALEKS test. In the same website, you will also be able to see the list of suggested topics for review.

Q: Where can I meet with an academic advisor?
A: Academic advisors are available at the MCC Enrollment Services Center, Southern and Dobson campus, Building 38. Enrollment Services at the Red Mountain Campus is located in

Source: MCCCD, 2010.

Appendix E:

ALEKS Progress Report

Pie Chart

Following the initial assessment, you receive a report in the form of a color-keyed pie chart. **In the ALEKS Pie, each "slice" corresponds to a particular area of the course,** such as "decimals" or "fractions and proportions." The darker portion of each pie slice represents the topics that you have mastered and the lighter portion represents what you have yet to learn.

The ALEKS Pie describes your current knowledge of the subject and makes precise recommendations for further study. Each slice of the pie chart may be opened to produce a list of the topics on which you can choose to work based on your prerequisite knowledge.

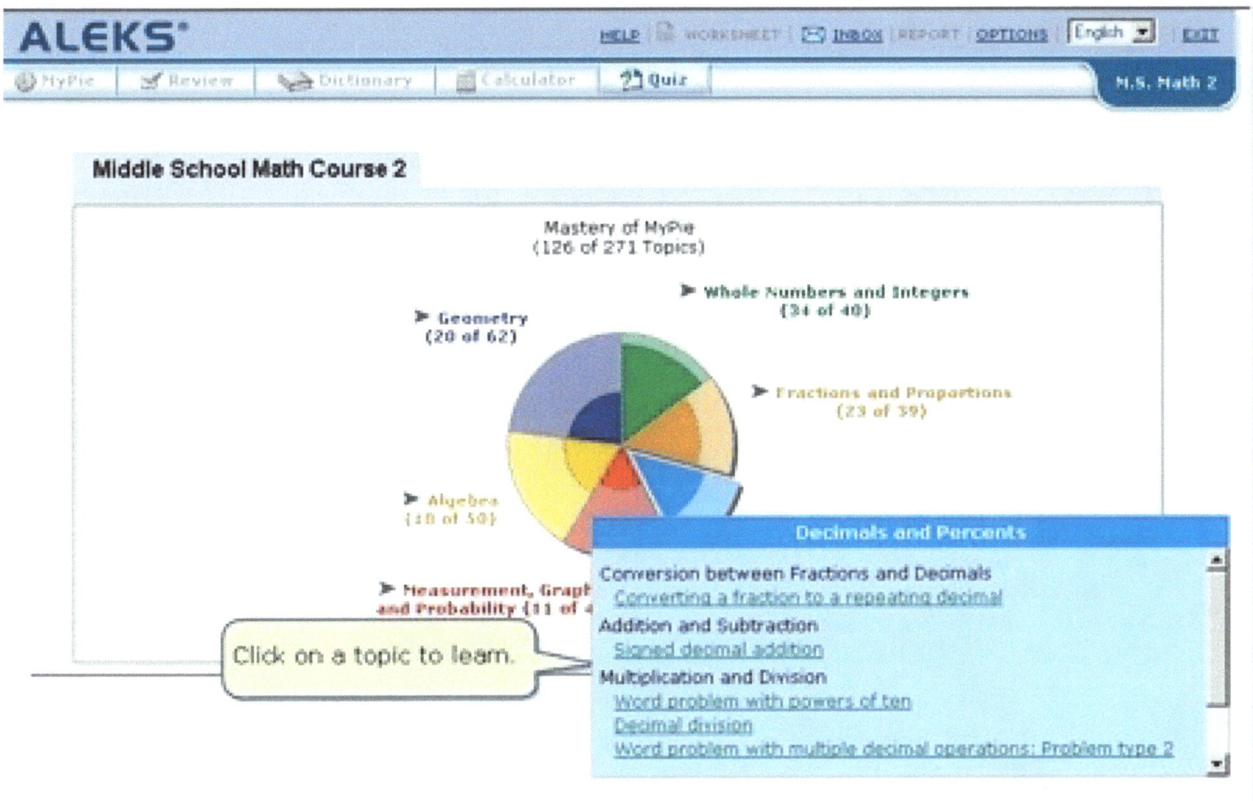

Detailed Progress Report

You can view a detailed progress report for yourself to determine what topics you know, what topics you do not know, and the topics you are ready to learn next. You can also view your learning progress history, as well as a learning log of suggested review topics.

ALEKS® Master Account

Inbox | Sign Out

Home | Edit Student Account | Reports ▾ | Quizzes ▾ | ALEKS QuickTables® ▾

Victoria Bolzano — Mathematics - LV 5 (with QuickTables)

Student Pie Chart

Print

05/13/2009 Learning

05/13/2009 (last login date) Learning

Total Hours: 35 hours 33 minutes

Course Mastery
(76 of 266 Topics)

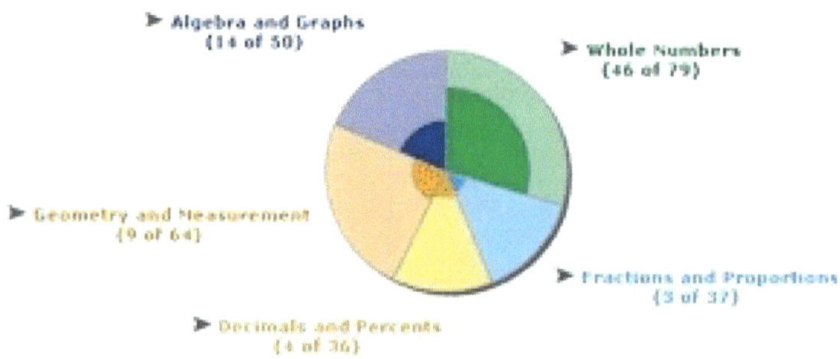

Victoria's State Standards Report

Change state to (Select a state)

What Victoria can do

Whole Numbers

Number Sense, Addition, and Subtraction
- Whole number place value: Problem type 2
- Expanded form with zeros
- Numeral translation: Problem type 2
- Word problem using addition or subtraction

Multiplication and Division
- Multiplication with trailing zeros: Problem type 1
- Using addition and multiplication to count the objects on a grid
- Division without carry
- Quotient and remainder: Problem type 1
- Basic word problem using multiplication or division

Rounding, Ordering, and Estimation
- Comparing a simple numerical expression with a number
- Rounding: Problem type 2

Algebra and Graphs

Data and Graphs
- Bar graphs for non-numerical data
- Line plots
- Interpreting pictograph tables

Probability
- Classifying likelihood
- Venn diagrams with two sets

Integers and Signed Numbers
- Plotting integers on a number line

Variables Expressions and Equations
- Evaluating a simple algebraic expression: Problem type 2
- Solving simple equations with multiplication or division

Functions and Graphs
- Describing data patterns
- Plotting a point in quadrant 1

and many other more elementary concepts.

What Victoria is ready to learn next

Whole Numbers

Number Sense, Addition, and Subtraction
- Subtraction with multiple regrouping steps

Multiplication and Division
- Multiplication with trailing zeros: Problem type 2
- Division with carry
- Word problem on quotient and remainder
- Word problem with multiple operations

Properties of Arithmetic, Fact Families, and Sequences
- Fact families for multiplication and division

Fractions and Proportions

Equivalent Fractions
- Understanding equivalent fractions
- Equivalent fractions

Ordering and Plotting Fractions
- Ordering fractions with same numerator

Multiplication and Division of Fractions
- Introduction to fraction multiplication

Decimals and Percents

Addition and Subtraction of Decimals
- Addition with money
- Decimal addition

Geometry and Measurement

Angles and Lines
- Measuring an angle with the protractor
- Introduction to parallel and perpendicular lines

Polygons and Circles
- Naming polygons
- Perimeter of a polygon
- Perimeter and area on a grid

Volumes and Surface Areas
- Classifying solids
- Nets of solids

Similarity, Congruence, and Transformations
- Introduction to congruence

History

Assessment type / Action	Date	Assessment performance goal	Time to completion (Current learning rate)			
			hours per week	items learned per hour of use	expected hours necessary to reach the goal	expected weeks necessary to reach the goal
Progress Assessment	05/09/2009	22 +7 %	7.7	1.6	117.2	15.2
Progress Assessment	05/02/2009	15 +7 %	11.1	1.0	114.7	10.3
Initial Assessment	04/19/2009	9 +6 %	5.4	1.9	114.7	21.1

Legend: Assessment ( ), progress in learning mode ( ), still to learn ( ), assessment not completed ( ), not assessed in this domain ( )

Learning log

No new topics have been learned since last assessment

Note: As a guest, changes you make will not be saved.

Source: ALEKS, 2010.

Appendix F:

F-Test Two-Sample for Variances for IT and RT

F-Test Two-Sample for Variances			
Descriptive Statistics			
VAR	Initial Test Score (IT)	Retake Test Score (RT)	
Sample size	172	172	
Mean	20.11	31.31	
Variance	75.02	271.43	
Standard Deviation	8.66	16.48	
Mean Standard Error	0.66	1.26	
Summary			
F	3.62	*F Critical value (5%)*	1.29
p-level 1-tailed	2.22E-16	*p-level 2-tailed*	4.44E-16
H0 (5%)?	*rejected*		

Appendix G:

Correlation and Regression Analysis for IT and RT Scores

Regression Statistics	
R	[illegible]
R Square	[illegible]
Adjusted R Square	[illegible]
S	[illegible]
Total number of observations	[illegible]

Correlation Coefficients Matrix			
Sample size	172	Critical value (5%)	[illegible]
		Initial Test Score (IT)	Retake Test Score ([illegible]
Initial Test Score (IT)	Pearson Correlation Coefficient	1	
	R Standard Error		
	t		
	p-value		
	H0 (5%)		
Retake Test Score (RT)	Pearson Correlation Coefficient	0.7	
	R Standard Error	0	
	t	12.8	
	p-value	0.00E+00	
	H0 (5%)	rejected	

Appendix H:

GLM Mixed Procedure Estimates for DIFFCPT by Demographics

The Mixed Procedure

Number of Observations	
Number of Observations Read	172
Number of Observations Used	163
Number of Observations Not Used	9

Iteration History

Iteration	Evaluations	-2 Res Log Like	Criterion
0	1	1188.92557491	
1	2	1174.10348444	0.00000116
2	1	1174.05797824	0.00000280
3	1	1174.05668830	0.00000000

Convergence criteria met but final hessian is not positive definite.

Covariance Parameter Estimates

Cov Parm	Group	Estimate
Residual	Ethnicity A	7.5922
Residual	Ethnicity B	241.67
Residual	Ethnicity C	29.3883
Residual	Ethnicity D	78.1975
Residual	Ethnicity E	131.82
Residual	Ethnicity P	1.4935

Note. Ethnicity A= American Indian/Alaska Native, Ethnicity B=Asian, Ethnicity C=Black/African American, Ethnicity D=Hispanic/Latino, Ethnicity E=White, Ethnicity P= Native Hawaiian/Other.

Appendix I:

Scatterplot IT vs RT

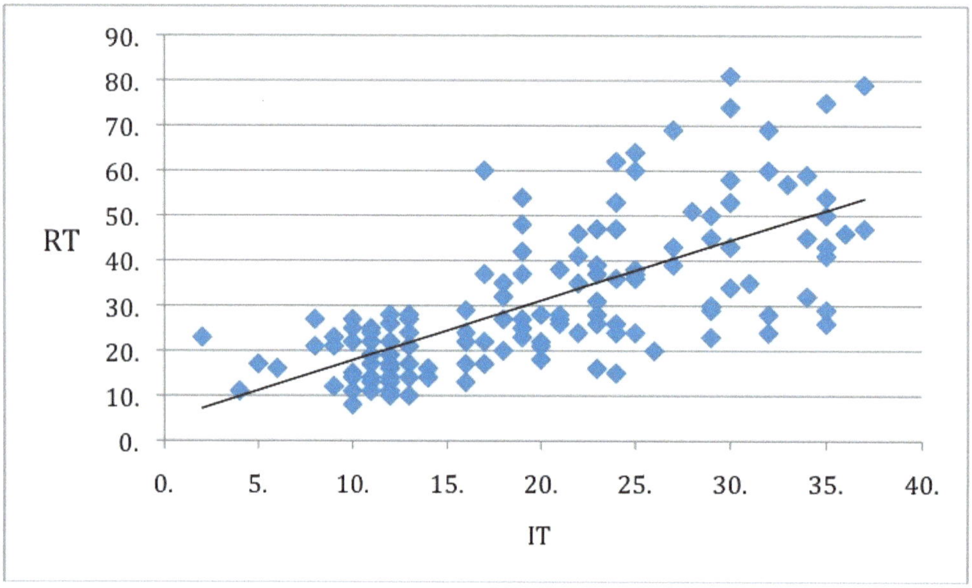

www.ingramcontent.com/pod-product-compliance
Lightning Source LLC
LaVergne TN
LVHW070936210926
848482LV00003B/249
* 9 7 8 0 9 7 6 7 3 0 5 6 9 *